AEROSPACE CLINICAL PSYCHO

Dedication

To my mother, Genevieve. I hope to prove that she made a wise investment when she rewarded me for finally paying attention in the tenth grade and for buying my monthly college train passes to New Brunswick, New Jersey.

And to my wife, Krystyna and my son, Elliott James, who have patiently endured missed birthdays and other special occasions as I pursued the experiences that have formed the basis for this book. Although I would like to pledge I will not miss these rites of passage in coming years, past promises have seemed to almost guarantee the next "opportunity to excel."

Aerospace Clinical Psychology

RAYMOND E. KING, Psy.D.

Routledge
Taylor & Francis Group

LONDON AND NEW YORK

First published 1999 by Ashgate Publishing

2 Park Square, Milton Park, Abingdon, Oxfordshire OX14 4RN
711 Third Avenue, New York, NY 10017

Routledge is an imprint of the Taylor & Francis Group, an informa business

First issued in paperback 2018

British Library Cataloguing in Publication Data
King, Raymond E.
    Aerospace clinical psychology. - (Studies in aviation
    psychology and human factors)
    1.Aviation psychology
    I.Title
    155.9'65

Library of Congress Cataloging-in-Publication Data
King, Raymond E., 1959-
        Aerospace clinical psychology / Raymond E. King.
            p.   cm. -- (Studies in aviation psychology and human factors)
        Includes bibliographical references.
        ISBN 978-0-7546-1105-9   (hbk.)
        1. Aviation psychology. I. Title. II. Series.
    RC1085.K53 1999
    155.9'65--dc21   99-30347
                                                                                    CIP

ISBN 13: 978-0-7546-1105-9 (hbk)
ISBN 13: 978-1-138-37823-0 (pbk)

Contents

Frontispiece

Capt King, we may take a bird during our flight. If I'm hit and I'm out, I want you to shake me and if I don't respond, point the aircraft at an uninhabited area and pull the handles. Don't worry, God will take care of me.

A senior aviator to the author at
the commencement of a T-37
"incentive" flight.

Preface

Although the rationale for this work is laid out in the first chapter, these acknowledgments may give the reader some insight into the origins of this book. During my formative years, I was under the tutelage of my self-taught mechanic grandfather, who not only (legally at the time) cut new treads in old tires, but even designed and made the tools to do so. He often provided lessons to me, once telling me that the most useful tool we own is our two hands, due to their almost infinite flexibility. This lesson was not lost on his audience, even though I was only about seven at the time. Earlier still, my maternal grandmother argued that it was best to not look at a cup of liquid I was unsuccessfully trying to carry. At the time I did not understand, nor could I be convinced of, the concept of operator-induced oscillation. My mother, who shared the profession of waitressing with her mother, consistently taught me the human factors lessons she had learned from carrying heavy and very hot plates. She even taught me how to carry four of these plates, without the use of a tray (a skill that serves me well at salad bars and buffets). The influence of my big brother, John Dennis King, was as pervasive as water to a fish. Surviving as his little brother taught me to be a respectful observer in a world ruled by those more powerful than me.

Immediately before starting my career, I was enthusiastically enumerating all the benefits of a stint in the Air Force to anyone who would listen. My focus, however, was redirected by Gene Nebel who challenged me with the remark: "Ray, you've mentioned all the Air Force is going to do for you, but what are you going to do for the Air Force?" With these words ringing in my ears, I set off for my career.

Soon I found myself recruited to be the next psychologist in a distinguished line to run an airsickness program with the colorful (now deceased) G. Kress Lochridge at Sheppard Air Force Base (Wichita Falls), Texas. He taught me to be proactive in my treatment of aviators and to value safety above all else. Alexis Hernandez shared all he knew with me and passed on the materials developed by the psychologists who came before us. Upon meeting the left-handed woman who would become my wife, everything in my life gradually became oriented exactly "wrong" (example: coffee pot handle facing the opposite direction). Shortly thereafter, I started eating lunch with the left-handed pilot/psychologist Walt Sipes who was always careful about where he sat (preferred: at the corner of the table with his left hand free). It wasn't until the birth of my son Elliott, however, that I truly learned about affordances as I spent all my "spare" time surveying our house to install countermeasures for all the things into which he could get.

John Patterson shared his interviewing skills with me and taught me to lecture (a skill I continue to hone). David R. Jones served as a supportive mentor who always made himself available, yet was never intrusive. Dick Hickman, Dick Jones, Terry Lyons, Ken Gliffort, Herminio Cuervo, Frank Carpenter, Bill Drew, and Joe Burton all served to tame the wilder side of my nature. Doug Ivan, on the other hand, served as my fearless advocate and taught me the value of fighting for that in which you believe. Meanwhile, Chris Flynn demonstrated the power of empirical research, particularly when it's funded. I still recall asking Chris why we actually needed funding to do research! Chris proved to be a great boss (after starting as a great colleague) and an enduring friend, in good times and bad. The team of Joe Callister and Paul Retzlaff demonstrated to me the sum is greater than the sum of its parts. Roy Marsh provided support to continue our work despite our geographic challenges. Gary Saboe consistently encouraged my research and writing proclivities.

Tommie Church and Murl Leibrect greatly encouraged and facilitated my continuing development as an aviation psychologist. Bob Helmreich and John Wilhem, both of the University of Texas at Austin, were always gracious and could find time in their busy schedules as I attempted to move from the pathological to the more normal realms of aviator functioning.

Before we pick up the story in Chapter One, I want to thank Ken Boff who proved to be a supporter as I continued my career at the Fitts Human Engineering Division, Wright-Patterson Air Force Base, Ohio after moving from the Ellingson Aerospace Medicine Consultation Service, Brooks Air Force Base, Texas. The influences at that heady institution, including my fortunate association with Dr. Boff's successor, Henk Ruck, and the intellectually stimulating discussions with Bob Eggleston and Mike McNeese, provided me with priceless learning experiences. The inspiration for this book occurred while I was preparing a presentation facilitated by Dr. Ruck, and enabled me to finish it. Similarly, Cindy Dominguez, Mike Vidulich, and John Flach, all connected with the Air Force Research Laboratory and Wright State University, generously shared their vast human factors knowledge with me. Becky Green consistently aided me in my quest for training and publishing; I am greatly in her debt.

Kelly Green and the Wright Brothers Chapter of the SAFE organization really got me to thinking that my brand of aviation psychology was something of interest to more than a half dozen individuals. I also want to acknowledge Ephi Morphew and all our colleagues in the Society for Human Performance in Extreme Environments. This small group of diversely talented professionals provides a welcome home for those of us who attempt to help others expand the envelope of human performance. Joyce Adkins has proven to be an enduring colleague who has helped me translate much of my clinical background into the broader context of organizational health.

The point of this convoluted beginning is to acknowledge the influences that led to this book. Each contributed to it, but I alone accept responsibility for any of its shortcomings. Please communicate with me at **SkyKing321@aol.com** to apprise me of whether this book has come close to hitting the mark. Last, but certainly not least, I wish to acknowledge the brave aviators whom I've had the honor and privilege to serve.

Case vignettes presented in *italics* are fictional portrayals of operators who contend with the challenges of hazardous environments. Although cases are not presented in a standardized format, each case includes the referral question, methods employed, and an emphasis on the aeromedical disposition. Later cases challenge the reader to formulate a disposition to illustrate that no single solution is ideal, with the policies and practices of the reader's nation and workplace imposing the standards of practice. The psychiatric diagnosis according to the various iterations of the Diagnostic and Statistical Manual are not included for several reasons: There has been inconsistency between various editions (particularly with regard to alcohol abuse and dependence) and the cases presented here were seen during the reigns of DSM-III, DSM-IIIR, and DSM-IV. Also, the aeromedical disposition of a case is often more clearly dependent on the regulations of the governing agency in force at the time the aviator is seen.

1   Is this Book Necessary?

Just what is "aerospace clinical psychology" and how is it different from clinical psychology? Why would the aviation and space realm need the services of a clinical psychologist anyway? Is this term just a blatant oxymoron?

Operational Definition

Aerospace clinical psychology is a special application of psychology to the hazardous and stressful occupations encompassed by aviation and extreme environments. Services are typically offered on a unit or organizational level but interventions can be tailored for individual operators and their families.

Current Status

Flight training and flying are hazardous activities that demand the most of human operators, whether they be pilots or other actors (maintainers, air traffic controllers, managers, regulators) involved in the complex aviation system. Clinical psychology is directly concerned with individual differences and performance under various conditions. Our current times have been termed "The Age of Anxiety." Nowhere is this felt more keenly than in the little-margin-for-error realm of aviation. Stress and exposure to risk are daily ingredients in the aviation business. Combining the talents of clinical psychologists with flight surgeons and human factors practitioners may enhance safety and efficiency.

Understanding, however, the strengths and vulnerabilities of most pilots can be a useful tool in rendering services to this extremely challenged, yet extraordinarily resilient, population. It is useful to understand the strengths and vulnerabilities of this group because when things start turning bad, they may do so in a major way. Unfortunately, the lore of aviators, delineated most elegantly by Dr. Frank Dully, suggests that they are unconnected to their emotional lives, and not introspective, they will have few or no mechanisms to deal with failure. If this characterization is accurate, the risk is that these so-called "failing aviators" will then be vulnerable to self-medicate or argue with spouse or convert their emotional conflicts to physical problems and present themselves to the flight surgeon with vague medical complaints. Before the flight surgeon orders the million-dollar work-up, it's good to know how these people are supposed to tick.

What is the current state of this proposed partnership? Moving from the clinical world to the human factors arena, one may feel like an anthropologist immersed into a different culture. Human factors practitioners include research and ecological psychologists, cognitive engineers, and, most puzzling of all, hybrids of the two. Early on, one of my new colleagues tried to explain *cognitive systems engineering*: "We look at the context of each situation, and assess it as a unique entity." I replied: "Just like clinical work." Later, I realized that "knowledge elicitation" is similar to clinical interviewing, but without, necessarily, an attempt at empathy. In other words, the skills of clinical psychologists closely map to those required of human factors practitioners. Nevertheless, I had an entire new vocabulary to learn, particularly when my boss kept criticizing one of my employees (and consequently me indirectly) for "going open loop." "Open loop" sounded as though it was a positive attribute, but I later learned (at one of our numerous "off sites") that an open loop doesn't allow for corrective feedback. A closed loop, on the other hand, is self-corrective (or in other words, in this case bosses are kept informed of what's going on).

Despite initial trepidation, the value of clinical training and aviation experience was demonstrated by requests for psychologists to contribute to aircraft safety (to prevent reoccurrence) and accident boards (to preserve evidence). The clinical psychology residents from the nearby medical center consistently amazed the staff at the SIRE (Synthetic Immersion Research Environment) facility due to their highly developed abilities with biofeedback. The self-regulation skills the SIRE researchers are attempting to teach operators are tightly coupled to those that clinicians attempt to teach patients.

"Human factors" are being recognized as a major contributor to aviation training attrition and the major cause of aircraft mishaps, and hence, represent a critical source of lost lives and flying resources.
Mental health care providers must be taught (and sometimes gently reminded) that the flight surgeon always remains the overall case manager, and only a flight surgeon can medically ground (DNIF - Duties Not Including Flying) an aviator. Mental health providers should only see

aviators on a consult basis from the flight surgeons office and should keep the flight surgeon informed of the aviator's progress (with the knowledge, of course, of the aviator). The relative lack of autonomy can sometimes be problematic for psychologists due to their history of struggling for recognition and a defined role in the delivery of health care. Nevertheless, psychologists may be uniquely suited to contributing to an understanding of "human factors" and can readily strike a workable balance between the needs of an individual versus the needs of the organization for which he/she works. The most productive relationships between flight surgeons and mental health providers are usually formed *before* an urgent need arises. In others words, the best referrals for mental health occur when the flight surgeon understands the services mental health can (and cannot) offer and has a firmly establish working relationship with the mental health provider. Beyond the needs of the immediate patient, mental health providers may make excellent safety presenters with a little encouragement and guidance from the kindly flight surgeon.

The Limits of Crew Resource Management

While a comprehensive discussion of the components of crew resource management (CRM) is beyond the scope of this book, an understanding of the concept of the "failing aviator," an aviator who is not adequately coping, may help managers and clinicians better understand why CRM itself sometimes seems to fail. While poor judgment is not a sufficient condition to warrant a mental health diagnosis, some aviators just don't seem to "get it." They may be able to get through CRM training (co-operate and graduate), but they don't incorporate its principles into their flying. These aviators may be situationally failing, due to increased life stressors, or may be chronic "screw ups," "accidents looking for a place to happen." The concept of "airmanship" is an important area to assess and includes the aviator's judgment, flight discipline, and commitment to safety. Experienced aviators and flight surgeons usually apply this test: "Would I fly with this person?" While the communication skills taught during CRM training may change attitudes, no amount of CRM training is likely to change personality traits, particularly maladaptive ones. An interaction from the clinician arena may be illustrative: A street-smart, psychiatrically hospitalized adolescent was confronted to make an "I statement" and to talk about feelings in an attempt to get him to own his affect and to be more available to other members of his therapeutic community. Without missing a beat he responded: "O.K., I hate you."

While he responded in the requested format, the intent of his message was exactly counter to what was desired. Similarly, teaching CRM skills to individuals who use words as weapons and view every social interaction as a zero sum gain with identifiable winners and losers will clearly not achieve the desired result: Increased safety and mission completion. Such individuals may be masters of catching and exposing the errors others make, but are not likely to be so vigilant of their own errors. I once had the good fortune to sit next to a submariner (or perhaps my seat mate had the poor fortune to sit next to a nosy psychologist) on a long commercial airline flight and was able to ask him how crews normally dealt with uncooperative and unpleasant crew members during long duration missions. He responded: "Those folks always seem to get themselves locked in storage closets where cans tend to fall on top of them." While this form of behavior modification may not be officially sanctioned, it clearly illustrates the ability of crews to adapt to what they're given.

The Future

Pilotless aircraft and advanced spacecraft lend unique challenges to the psyche of the operator, as does rapid change from localized flare-ups to global nuclear threats. Experts in psychological research will be tasked to help aviators and policy makers keep the operator up with the rapid changes. As we invest increasingly large amounts of money into each individual airframe and mission, we must learn more about the human operator, whether that individual is a pilot or an operator in a virtual reality environment.

Complex systems demand selection and training research to ensure compatibility of future operators and highly advanced technology. These areas are even more important when several other changes are considered, including: uninhabited aerial vehicles, fewer pilots and aircraft to accomplish the mission, greater numbers of female aviators, increased co-operation between nations, the need for collaboration in complex systems, and an ever-changing threat.

Today's, and tomorrow's, successful systems operator may be very different from the barnstormers and aces of yesteryear. In particular, military aviation has changed from the days of dog fighting to modern multi-crew, highly coordinated missions. Due to ever-increasing levels of complexity, future military and space operations will be more highly dependent on team functioning.

Aviation, particularly military aviation, is increasingly an interpersonal endeavor. A study of the psychological factors leading to success in aviation will enable researchers to select-in the most psychologically and cognitively suited person for the mission rather than only select-out those who are inappropriate. The need is rapidly changing from individual excellence to group excellence, which may be more than the mere sum of the characteristics of individual members of a team. Even individuals who have done well in solitary pursuits may fail when placed in an interpersonally demanding situation. Space exploration serves as a useful model. Due to lengthening space missions and the increasingly interpersonal nature of space exploration, combined with an expanding multicultural flavor, selection of individuals with desirable interpersonal qualities is becoming paramount. The basic question in the back of the minds of evaluators for the space program is: "Could I spend six months in a bathroom with this person?"

United States Air Force pilots, and pilots from other services and nations, are selected on instruments heavily loaded on general intellectual ability (Carretta, Retzlaff, Callister, and King, 1998). Special programs often require additional selection methods, such as psychological testing and structured interviews. Moreover, the psychiatric standard of fitness, which is an intrapsychic phenomenon, is yielding to a psychological model of suitability, which is more context specific.

The air forces of the future will surely include more women and they will likely compete on an equal footing and may be represented in all cockpits. The operators of the future will face an ever-changing enemy. As nation-states and political systems rise and fall so to will the nature of warfare and war machines. The cognitive abilities and personality make-up of combatants may need to change with both the enemy and technology. Pilotless aircraft and advanced spacecraft lend unique challenges to the psyche of the operator, as does rapid change from localized flare-ups to global nuclear threats. Experts in psychological research may be tasked to help aviators and policy makers keep the operator up with the rapid changes. As we invest increasingly large amounts of money into each individual airframe and mission, we may also benefit from learning more about the human operator, whether that individual is a pilot or an operator in a virtual reality environment.

Future research should consider cultural differences and expectations regarding the roles of men and women in a multinational collaborative effort. The gender and cultural make-up of a unit and the mission they are tasked to accomplish may affect crew coordination, squadron relationships, mission effectiveness, and flight safety.

2 Selection: What Can a Clinically Trained Psychologist Contribute?

Select-out *vs.* Select-in

Much to the surprise of many of our international allies, clinical psychologists play no direct role in the selection of pilots in the USAF. Behavioral scientists, however, have a long record of working on research questions and providing practical services in the USAF. So practical that at the time of this writing, selection does not seem to have a place in the newly reorganized USAF Research Laboratory as the Laboratory Commander views selection work as more of a service than a scientific endeavor.

What do clinically oriented, as opposed to their research-oriented colleagues, contribute to selection? USAF flight surgeons are directed to conduct a semi-structured interview, termed the Adaptability Rating for Military Aeronautics (or Aviation, ARMA), to assess motivation to fly and solicit limited biographical screening during the initial flight physical. The flight surgeons ask questions about "aviation affinity" (why candidates want to be a pilot) as a brief and crude way to detect unsuitable candidates (Mills & Jones, 1984). The ARMA, however, is inconsistently used and flight surgeons are not satisfied with it (Verdone, Sipes, & Miles, 1993). In the case of returning a trained (termed "rated" in the USAF, "designated" in the US Navy) pilot to flying duty, Adams and Jones (1987) propose a professional interview as the best way to assess very subtle factors in this typically healthy, well-defended population. Adams and Jones explain that grounded flyers are usually intelligent, articulate, and eager to resume flying duties while also being "rarely attuned or introspective, making them particularly vulnerable to the psychosomatic manifestations of anxiety" (*p.* 350).

Currently, US Air Force clinical psychologists, and other clinical psychologists who assess aviators, use standardized personality measures such as the Minnesota Multiphasic Personality Inventory-2 (MMPI-2) when evaluating referred pilots. These tests have been normed on the general population, and several studies (Ashman and Telfer, 1983; Picano, 1991; Retzlaff and Gibertini, 1987) demonstrate that pilots, as a group, differ from the general population. For this reason experienced aviation psychologists use pilot-based normative data whenever possible. However, appropriate pilot norms are difficult to establish since psychological tests are rarely given to large representative samples of pilots.

The USAF psychologically tests aircrew members when seeking to return them to flying status. Formerly, the psychological testing of these medically referred members were compared to MMPI results complied during the selection for the Apollo (space) program (Fine & Hartman, 1968). Other psychological tests may be problematic for the aviator population as many pilots would be classified as having a personality disorder on the basis of their psychological testing (King, 1994). These tests are therefore not used or are used with great caution.

Neuropsychiatrically Enhanced Flight Screening (N-EFS)

The Enhanced Flight Screening program is a short flying program originally intended to help reduce elimination from USAF Undergraduate Pilot Training (UPT). The Neuropsychiatrically Enhanced Flight Screening (N-EFS; King and Flynn, 1995) program was a response to the USAF Surgeon General desire to identify pilots who would be likely to be later medically disqualified. Since such a task was not possible with current psychological norms, a psychological baselining and research program was needed. Another driver for the N-EFS program was commanders' desire to identify who would make the best pilots. Now all candidates for USAF pilot training participate prior to entering the Enhanced Flight Screening Programs at Hondo, TX and the US Air Force Academy in Colorado Springs, CO.

N-EFS, established in 1994, archives baseline intelligence (using the Multidimensional Aptitude Battery, MAB; Jackson, 1984) and cognitive functioning (using the CogScreen; Kay, 1995) for comparison purposes if a future medical waiver to aircrew standards is needed, post head-injury. Some pilots will sustain head injuries that affect their flying careers. The complexity and unforgiving nature of their working environment demand a conservative approach to occupational return after neurological insult. Therefore, a medical evaluation is required to return to flying duties, possibly including neuropsychological assessment. Lacking pre-injury neuropsychological data, however, accurate assessment of post-injury functioning is hindered. Baseline intellectual and cognitive data support the scientific basis of a waiver for return to flying status.

Multidimensional Aptitude Battery (MAB):

Verbal IQ - Crystallized ability (results from interaction with the culture)

- *Information* - Fund of knowledge, long-term memory
- *Comprehension* - Ability to evaluate social behavior
- *Arithmetic* - Reasoning and problem-solving ability
- *Similarities* - Flexibility, adjustment to novelty, abstract thought, long-term memory
- *Vocabulary* - Openness to new information, capacity to store, categorize and retrieve words and verbal concepts previously learned

Performance IQ - "Fluid" ability (independent of education and experience, capacity for learning and problem solving)

- *Digit Symbol* - Adaptation to a new set of demands, learning coding, performing visual-motor tasks
- *Picture Completion* - Identifying important missing elements in a picture, knowledge of common objects
- *Spatial* - Ability to visualize abstract objects in different positions
- *Picture Completion* - Ability to identify a meaningful sequence, social intelligence and insight into others' behavior
- *Object Assembly* - Visualization skills and perceptual analytical skills needed to identify a meaningful object from a left-to-right sequence

Full Scale IQ - General aptitude (compilation of all subtests).

Note: IQ scores are standardized scores with a mean set at 100 and a standard deviation of 15.

As a group, N-EFS students are scoring in the superior range on baseline intellectual assessment, but with considerable variability (spanning below average to very superior). This wide range of intellectual functioning in these pilot candidates argues for baseline data collection that optimizes future waiver decisions. These assessments would compare the aviator's post-injury functioning to a personal intellectual functioning baseline captured at entry into aviation training. Personality testing suggests few significant differences between male and female student pilots, with high extraversion being the most striking feature, particularly among males (King and Flynn, 1995). The challenge of evaluating an aviator without benefit of baseline data is demonstrated by the following fictional case vignettes.

Case Vignette:

Mr. Gump, an athletic pilot with remarkable dexterity, sustains a mild/moderate head injury. Upon presenting himself for neuropsychological evaluation, his Full Scale IQ is found to be 95. Believing this result to be at wide variance from his (unmeasured) baseline functioning (average aviator IQ is 120.06 with a standard deviation of approximately 6.72; Carretta, Retzlaff, Callister, and King, 1998), the psychologist recommends continued medical grounding.

It is possible that our referred aviator suffered no reduction in his cognitive functioning as a result of this injury and was a successful aviator with an IQ in the average range. We need to compare individual aviators to themselves. While there are few successful aviators with IQs "only" in the "normal"* (90 - 110) range (King and Flynn, 1995), it is best to know an individual aviator's baseline information. That way we can perform an ideographic, as opposed to a nomenthetic, assessment.
(* "Normal" based on general population norms; Jackson, 1984)

Case Vignette:

A medical doctor calls a psychologist to seek guidance about a helicopter pilot in his mid sixties. This aviator's supervisor is concerned with a recent onset of critical omissions, almost resulting in tragic consequences on more than one occasion. This aviator, however, continues to successfully negotiate check rides and routine flights. The flight surgeon wants to send him to a local psychologist. You advise against this course of action due to your concern that this aviator would be compared against his age cohorts. Subtle impairments could easily go undetected. You instead advise referral to the tertiary facility where at least a baseline assessment could be established to gauge any dementing process.

Psychological data may also improve the aviator selection process as financial resources dwindle. N-EFS is therefore also attempting to validate the MAB, CogScreen, Revised NEO-Personality Inventory (NEO-PI-R; Costa and McCrae, 1992), and Armstrong Laboratory Aviator Personality Survey (ALAPS; Retzlaff, Callister, and King, 1997) for pilot selection and assignment to specific cockpits. Participation in this part of the program is completely optional for UPT candidates.

Goals of the Neuropsychiatry Portion of EFS

- Collect Baseline Data (Medical - mandatory).
- Establish Pilot Norms (Research - optional, informed consent solicited).

Factors Tested

- Intelligence (Medical).
- Cognitive Skills (Medical).
- Personality Characteristics (Research) - which includes administration of the NEO-PI-R and the ALAPS.

The NEO-PI-R is a test designed to measure normal personality characteristics. It consists of 240 statements to which the evaluee responds on a scale from 1 to 5, which represents "strongly disagree," "disagree," "neutral," "agree," or "strongly agree." Scored NEO-PI-R's provide five domain scores.

NEO Personality Inventory-Revised (NEO-PI-R)

- *Neuroticism (N)* - Level of emotional stability.
- *Extraversion (E)*- Sociability, assertiveness, activity.
- *Openness to Experience (O)* - Imagination, aesthetic sensitivity, attentiveness to inner feelings, preference for variety, intellectual curiosity, and independence of judgment.

- *Agreeableness (A)* - Altruism, sympathetic to others and eagerness to help, belief that others will be sympathetic.
- *Conscientiousness (C)* - Self-control, determination.

Armstrong Laboratory Personality Inventory (ALAPS)

To address the need for a more crew-oriented instrument, we developed the ALAPS as we determined the need for a new test that measured pilot specific personality characteristics. We determined the specific domain areas to be tested and generated large pools of items for these domains. Raw scores can be converted to standard T scores, with a mean of 50 and a standard deviation of 10.

Armstrong Laboratory Aviator Personality Survey (ALAPS)

Personality Scales
- *Confidence (CO)*:   High scorers view themselves as highly capable, intelligent, and talented. This tendency may include the negative elements of arrogance, manipulation, and condescension. Clinically, these traits may suggest narcissism.

- *Socialness (SO)*:   High scorers are extremely social and outgoing. They enjoy others and are socially comfortable. They see themselves as friendly and charming. Clinically, these traits may include elements of histrionic personality disorder.

- *Aggressiveness (AG)*: High scores are assertive to the point of being aggressive. They take strong stands and tolerate little criticism. They are verbally and emotionally combative. This quality probably does not rise to the level of antisocial personality disorder.

- *Orderliness (ORD)*: High scorers are orderly in a behavioral and environmental way. Their lives are structured and neat. They are methodical and disciplined. Clinically, this tendency may rise to the level of compulsive personality disorder.

- *Negativity (NE)*: High scorers are angry, negative, and cynical. They are socially punitive and not pleasant to be around.

Clinically these traits may rise to the level of negativistic or passive-aggressive personality disorder.

Psychopathology Scales
- *Affective Lability (AF)*: High scorers are generally emotional and reactive. They are situationally anxious, depressed, and frightened. Moods are seen as changing quickly with little provocation. Affect is volatile.

- *Anxiety (AN)*: High scorers are chronically anxious. They worry and brood. The anxiety interferes with their lives and occupational functioning.

- *Depression (DE)*: High scorers are depressed. Problems include dysphoric affect as well as the cognitive and vegetative symptoms of depression. They report being pessimistic, unhappy, and guilty. Extreme elevation may include clinical major depression.

- *Alcohol Abuse (AL)*: High scorers like to drink, drink a great deal, and get intoxicated. Functioning is impaired and there may be social and occupational problems.

Crew Interaction Scales
- *Deference*: High scorers are deferent to a fault. They are submissive and quiet. They concentrate on their job and are not likely to question the status quo.

- *Dogmatism*: High scores believe they are always correct and are not open to change. They are interpersonally authoritarian. They are intolerant of other people, their ideas and their actions.

- *Impulsivity*: High scorers act first and then think. They often act and talk without sufficient forethought. They see themselves as "spontaneous."

- *Organization*: High scorers are systematic and organized. They coordinate and plan all elements of a project. They thoroughly think things through.

- *Risk Taking*: High scorers enjoy danger and risk. New activities and situation are not of concern. They are adventurous, unafraid, and fun loving. They are not necessarily impulsive about their activities; their action may be calculated and include a rational appreciation of the inherent danger.

- *Team Oriented*: High scorers enjoy and believe in teamwork. They value team effort and team rewards. They do not enjoy working alone and may be inefficient when doing so.

Previous Efforts

Calls for better selection and screening techniques are older than the USAF (War Department, 1940). Choosing among applicants for aviation duty from a pool of high-functioning, accomplished individuals is a difficult task. Select-out criteria, or eliminating applicants with a psychiatric diagnosis (lack of fitness), results in the identification of a very small impaired subset of the candidate pool, but does not suggest who is the most qualified or adaptable applicant. Select-in methods determine who is best suited for challenging tasks, but does not screen well for psychopathology. Psychometric data on successful aviators and other professionals, particularly those who fly aboard aircraft requiring high degrees of crew co-ordination, are historically very limited. The NEO-PI-R measures the normal range of personality functioning and can be used as a select-in measure to determine suitability, or as select-out measure to determine fitness. The NEO-PI, the predecessor of the NEO-PI-R, was researched, at least in part, as a select-in measure for the 1989 selection cycle (Santy, 1994). The NEO-PI was not used during the author's involvement in the 1991/1992 NASA astronaut selection cycle. Instead, standard select-out tools, such as the MAB, the MMPI (now revised to be the MMPI-2), the Millon Clinical Multiaxial Inventory-II (MCMI-II), the Forer Sentence Completion technique, a family history questionnaire, and structured clinical interviews were employed to determine if an applicant had a disqualifying psychiatric diagnosis. Other behavioral scientists invited NASA astronaut applicants to volunteer for a research study (select-in) using the Personal Characteristics Inventory (PCI; Chidester, Helmreich, Gregorich, and Geis, 1991), a test of judgment and potential for effective crew resource management. The NEO-PI-R was used as a select-out tool during the 1994 NASA astronaut selection cycle, replacing the Millon Clinical Multiaxial Inventory-II, which research (King, 1994) suggests may exaggerate psychopathology, particularly in a non-referred population. Our job was to determine if an applicant had a disqualifying psychiatric diagnosis. Other behavioral scientists (psychiatrists/ psychologists) were present to invite applicants to volunteer for a research study (select-in) using the PCI. All applicants were offered feedback (of about 15 minutes) on their clinical testing (select-out) and clinical interview performance. Subsequent selection cycles included a combined select-in/select-out assessment, with consideration given to the likely psychological requirements for long duration (International Space Station) missions.

Previous studies of military aviation training typically attempted to use the criterion of success as completion of training, rather than actual mission readiness. Such short-term research is frequently plagued by the "honeymoon effect;" students, attempting to look their best, can sustain a high level of performance in the short-run (Helmreich, Sawin, and Carsrud, 1986). While many measures can discern who *can* (aptitude) finish pilot training, there is scant information available on who *will* (motivation) finish pilot training and become effective aviators.

A good predictor of success in flight training has traditionally been previous flight hours. The value of previous flight hours, however, may actually be its indication of the individual's history of motivation to fly. In other words, seeking out flight training speaks well of the depth of one's enduring motivation to fly. Many applicants for astronaut candidacy seem to present themselves with a few flight hours, hastily accumulated in the months immediately before their interview. A significant accumulation of flight hours is a good indication of solid motivation to fly.

Cross-Sectional Study of Female (and Male) Pilots

The determination of psychological fitness to fly is complicated, particularly when attempting to extrapolate what little we know about male aviators to women. The United States rescinded the combat exclusion clause and consequently, the USAF now officially allows women to potentially engage in aerial combat. The Defense Women's Health Research Program (DWHRP), funded by the United States Congress, allowed researchers to study the needs of women in military service. "Assessment of Psychological Factors in Aviators" (APFA), supported by the DWHRP, studied successful incumbent pilots. APFA did not include the very few females flying fighter aircraft. Nevertheless, many of the participants vigorously maintained they were exposed to enemy fire while flying during the Persian Gulf War. They did not, however, have the opportunity to return fire. Furthermore, they did not have the relative peace of mind of sitting in an ejection seat; often they were also flying with thousands of pounds of fuel to refuel other aircraft. In contrast to the longitudinal approach of N-EFS, APFA was cross-sectional. Another difference between N-EFS and APFA was that we also interviewed participants with a semi-structured interview during APFA.

The US Army Aviation Flight Program opened to women in 1973. On 7 October 1975, US President Gerald Ford cleared the way for women to enter the service academies. Women began training as pilots and navigators in the USAF in 1976 and became US Navy flight officers in 1979. The Department of Defense lifted the Combat Exclusion Clause in 1993, clearing the way for women to potentially fly in aerial combat.

With increasing numbers of female military pilots, the USAF deemed it important to understand the psychological and psychiatric gender differences of pilots. Using the NEO Five Factor Inventory (NEO-FFI; Costa and McCrae, 1992), female United States Air Force pilots were compared to both male Air Force pilots and to a female comparison group. The NEO-FFI, similar to the NEO-PI-R, is a survey of the normal range of personality functioning but is shorter as it consists of 60 items. The five domains of neuroticism, extraversion, openness to experience, agreeableness, and conscientiousness are gauged, but without the more detailed facets.

Female pilots may bring different personality styles into the cockpit and understanding those differences is important both medically and operationally. We solicited data within flying squadrons from non-medically referred USAF pilot volunteers. The purpose of the study was to compile psychometric norms on both female and male pilots.

Women currently flying United States Air Force (USAF) aircraft were selected using methods developed for men (Siem, 1990). The structure of the paradigm of the "Right Stuff" (Wolfe, 1980) rests on a male foundation. Do female pilots bring different intellectual skills and personality styles into the cockpit? APFA solicited data within flying squadrons from non-medically referred USAF pilot volunteers. One hundred and fourteen (50 women, 64 men) fully qualified USAF pilots currently assigned to crewed aircraft took the MAB, the intelligence (IQ) test described in a previous charter. The MAB is presented in a multiple choice format as 10 seven-minute subtests. The Armstrong Laboratory helped computerize its administration; this version equates well with the paper-and-pencil version (Retzlaff, King, and Callister, 1995a).

MAB results of the APFA volunteers revealed no significant differences in intellectual skills, while a comparison of female pilot versus male pilot personality structure suggests greater female extraversion, agreeableness, and conscientiousness (King, McGlohn, and Retzlaff, 1997). These traits may be highly adaptive for Air Force pilots, given the nature of modern military operational requirements.

The observed lack of intellectual differences, despite the general population pattern of male versus female disparity (Halpern, 1992), may be a function of selection (both self-selection and military personnel selection) and assignment. Conversely, these female pilots seem to have even more positive personality traits. Lyons (1991) notes that all aviators have a unique psychological profile and female pilot candidates are not a representative sample of the general population. Hence, determination of *female* psychological fitness to fly is complicated. Moreover, Jones (1983) cautions flight surgeons about the particular challenges, such as being cast into various inappropriate roles by their male brethren, faced by female aviators.

On interview, the majority of male pilots voiced concern about their proclivity to protect women in combat. Female participants were concerned about potentially being used to exploit captive male comrades-in-arms, thus raising an important training issue (McGlohn, King, Butler, and Retzlaff, 1997)

Opportunities for further study include directly comparing applicants and candidates for military flying training to incumbent pilots to address the aviation nature versus nurture argument. We need to discern whether pilot training *changes* women to be more like men or whether certain women self-select themselves into pilot training.

3 The "Myth" of Pilot Personality

The Right Stuff

The mythical paradigm of the "right stuff" for military pilots holds that these individuals possess extreme levels of confidence, assertiveness, and competitiveness. The typical lay impression holds that all military pilots have a singular personality type. While this idea may make for good Hollywood movies, it does not make for good science.

The study of the personality characteristics of pilots has a long and controversial history. Psychologists first measured pilots' personality characteristics during World War I, and even at that time there were starkly divergent ideas about which personality characteristics were most important. For example, Rippon and Manuel (1918) described the ideal pilot as high-spirited and happy-go-lucky, while, three years later, Dockeray and Isaacs (1921) described the ideal pilot as quiet and methodical. The controversy over pilot personality continues today, driven primarily by strong evidence that personality measures are poor predictors of completion of initial training (Hunter & Burke, 1995). On the other hand, personality measures may have more utility in predicting performance beyond initial training completion. For example, Houston (1988) found that personality measures were the best predictors of the ratings given to first officers by captains. And, personality measures taken during initial training appear to predict retention characteristics in US Air Force pilots (Retzlaff, King, & Callister, 1995b).

Despite the controversy over the relationship of "normal" personality characteristics with pilot performance, there is little argument that there are "abnormal" personality characteristics that are undesirable. Highly anxious, hostile, or impulsive people probably should not control aircraft. In the United States Air Force, personality disorders are not medically disqualifying; however, administrative separation can occur when personality characteristics are judged to significantly impair the performance of military duties.

There seem to be differences within pilot samples and there is not one personality style among all pilots. Retzlaff and Gibertini (1987) found three distinct personality types among student USAF pilots. Only one of these three personality types seemed consistent with the "right stuff" lore. Similarly, Picano (1991) studied US Army pilots and concluded there is no single successful pilot personality dimension. Picano further found experienced pilots to be distributed among three personality clusters, with the "right stuff" type represented by only 16 per cent of his sample. The largest is the achievement-oriented, dominant, and affiliative type that takes a practical approach to problem solving. This subgroup tends to be level-headed and value officer and camaraderie. Another subgroup is similar to the first but tends to be more dominant, aggressive, exhibitionistic, and self-aggrandizing. These are the pilots with the "right stuff." The third and final subgroup is cautious, compulsive, and socially retiring. This subgroup tends to be the least socially affiliative or achievement-oriented. Membership to any of these groups does not strongly predict success or failure in military aviation.

These studies as well as other collections of psychometric norms were invariably based upon male pilot subjects. Additionally, the subject pools were often severely limited or based on very small, specialized populations such as fighter pilots, test pilots, or astronauts. Data on female aviators had been scarce.

Novello and Youssef (1974) found general-aviation female pilots to be more similar to their male pilot counterparts than to women in the population at large. Specifically, they found female pilots deviate in the same direction as their male counterparts on 15 of 16 Edwards Personality Preference Schedule scales. Novello and Youssef hypothesize pilot personality styles that transcend gender.

The interest in aviation circles about personality is directly related to selection, training, performance, and safety concerns, as well as mental health considerations. Siem and Murray (1994) found experienced pilots rated "conscientiousness" as the most important of the "big five" (neuroticism, extraversion, openness to experience, agreeableness, and conscientiousness) personality characteristics determining pilot performance. Siem and Murray, therefore, advocate research to further validate the importance of conscientiousness in pilot personality and performance.

There's an old saying: "You can always tell a fighter pilot, but you can't tell him much!" Tom Wolfe (1980) describes the "right stuff" as an amalgam of stamina, guts, fast neural synapses, and old-fashioned hell raising. The popular media has saturated us with other books and movies that present us with profiles of pilots. Is there, however, just one type of pilot? Anyone who says that the book and subsequent movie, *The Right Stuff* (Wolfe, 1980), presents a mere stereotype did not pay close enough attention to the differences in the personalities of the original seven Mercury astronauts.

Major Flaws to Pilot Personality Theory

- Pilots are not clones of each other. Remember, "modal" refers to "most frequently occurring." Similar to snowflakes, no two pilots will ever be exactly the same.

- Fighter, test, and light attack pilots along with astronauts are the most frequently studied, at the expense of tanker/transport pilots, navigators, weapons officers, flight surgeons, and enlisted flying personnel. Most times the information that applies to one group is extrapolated to the other groups regardless of the actual validity of doing so.

Remember also, there are many different motivations for flying. So, therefore, attempt to understand the individual aviator. Avoid preconceived notions or stereotyped, cookie-cutter caricatures.

The "Modal" Aviator

Let's now look at the typical aviator (based on the observations of Dr. Frank Dully, retired U.S. Navy flight surgeon):

Modal pilot characteristics

- The pilot is a controller! That which cannot be controlled is avoided, at all costs. Situations are typically not entered into without there being a clear plan for egress or ejection. The concept of control not limited to controlling only an aircraft: Everything must be controlled (spouse, children, dog, car, house). If the controller is not in control or can't at least pretend to be, he or she will be very irritable. Controllers hate surprises, so they practice, practice, and then practice some more. That is what "Red Flags" are all about (chance to practice war). These individuals plan their spontaneity! How are these individuals likely to do with emotions? Male pilots usually also marry first born, controlling women. Trouble brews after he returns from extended deployment.

- Emotional distance from others is maintained at all times. These folks have difficulty with intimacy in their marriages. That's why temporary duty was invented. Male pilots often list "communication" as deficient in their marriage and in need of improvement or are completely unaware of any problems at all in there marriage and taken by surprise when wife (or husband) leaves. On psychological testing, this group will typically score high in gregariousness but low in warmth; they are extroverted introverts. Remember, this group is high achieving, it's difficult to nurture friendships while studying.   This group typically majored in science or engineering and not liberal arts in college.

- The pilot is systematic and methodical. These people rely on checklists and feedback. Their goal is to avoid surprises so they are likely to appear inflexible or even rigid.

- Pilots have the ability to separate flying and nonflying-related issues so that they may be dealt with at the appropriate times. They are mission-oriented compartmentalizers. If an issue is not connected to the mission at hand, then it is ignored. It is no surprise then these individuals are unconnected to their emotional lives.

In consideration of the above, do you think that a pilot would be a fun person to take on a vacation? Practitioners who work with pilots should be aware of stress periods when the aviator's skills may be overstressed:

- Birth or impending birth of child (particularly due to fatigue).
- Marital problems.
- Serious personal illness (whether or not a grounding condition).
- Base closure and/or loss of flying job.
- Referral for psychological evaluation.

The Generic Pilot

- First born or only son, or in the case of many female aviators, took the place of the son that never was. All seven of the original Mercury astronauts were first-born sons. Twenty-one (21) of the first 23 astronauts to make space flights were first-born sons (Reinhart, 1970). If the aviator is not the first born, then he is the functional equivalent due to some unacceptable quality of the chronologically first born.

- Whether male or female, unusually close to father (who is often a pilot himself).

- Regardless of gender:
  Adventurous
  Courageous
  Skillful
  Competent
  Masters of complex tasks
  Self-confident but easygoing
  Little desire for psychological insight
  Need for autonomy
  Practical logician
  Heterosexual (Female pilots are more like male pilots than they are like females in the general population. They, too, however, are decidedly heterosexual.)
  Extremely physically and psychologically healthy
  Alloplastic (rather change environment than change self)
  Self-sufficient
  Short-term goals preferred over long-range planning
  High achievement need
  Inflexible
  Seeks out novelty and responsibility
  Interpersonally Direct
  Above average IQ (Average pilot IQ equals 120, placing them in the 90th - 95th percentile for intelligence, superior range)
  Energetic
  Low tolerance for imperfections

In sum, flying satisfies their need for achievement, individual initiative, novelty, excitement, and responsibility.

We don't know if the characteristics peculiar to the pilot are required for the task, or if the job attracts people with these traits, or if both factors are operative. The generalizations about pilots may also be true of other demanding professions (particularly first-born finding). The difference between a successful aviator and the aviator who does not do as well is often slight. Even when a pilot "fails," there is likely to be an absence of underlying psychopathology.

Contradictions in the Make-up of the Pilot that May Make the Concept of "Pilot Personality" Appear Less Valid

- High intelligence but non-intellectually oriented.

- Team player but anxiety in close relationships. Likely to be gregarious and appear extroverted but in reality not particularly warm. Many acquaintances but very few, if any, close friends.

- Appear easygoing despite driven quality.

The Coping Pilot as Compared to the "Failing Aviator" (Not-So-Coping Pilot)

Coping Pilot	Not-so-coping Pilot
•High mastery need	•Aggressive[1]
•Type A personality who is highly driven	•Also Type A, but impulsive/ impatient
•Exhibits high performance	•Exhibits low performance
•No health problems	•Vague health problems
•Good cockpit decision making and management skills (crew resource management)	•Questionable decisions, so-so or uneven management skills

•Has discipline and good judgment	•Impulsive. Lacks judgment (takes unnecessary risks because is trying to prove that he is not scared; counterphobic)
•Knows his (her) capabilities and limits. Knows aircraft's capabilities and takes it to its limits. (In the case of a test pilot the envelope is larger)	•Attempts to live out the fantasy of the "hot" pilot as depicted in movies. Takes the aircraft well beyond its capabilities
•A professional and a team player	•Egocentric (thinks only of himself)

Final Danger

There will be many similarities between pilots and physicians (sometimes they are the same person). There is a danger, therefore, of losing objectively. Non-pilot health care providers need to be aware of any "wannabe" (wanting to be a pilot) tendencies.

One difference: Flight surgeons score higher on "nurturance" (to take care of others) while pilots score higher on "dominance" (to insist on being the leader) on the Edwards Personal Preference Scale (EPPS).

Note

1   An obviously agitated member of a flight surgeon audience once took exception to this distinction, arguing that the speaker did not have any idea of what he was talking about as there is no difference between an individual with "high mastery need" and one who is "aggressive." This flight surgeon demanded to know the difference and challenged the speaker that there was, in fact, no distinction. The speaker replied: the difference can be appreciated by considering the questioner who seeks to truly understand versus the questioner who wishes only to humiliate the speaker.

4 Getting Information: Psychological Testing, Interviewing, Other Data Gathering

Psychological Testing

Psychological testing is the one activity that distinguishes psychologists from their psychiatrist and social worker colleagues. Curiously, psychological testing is one of the least preferred, and to an extent consequently least performed, activities of a clinical psychologist. To many, psychological testing is mysterious, perhaps almost verging on the mythical. What, in fact, is psychological testing? Psychological testing is merely a sample of behavior under standardized conditions used to predict or infer other, more relevant, behaviors. The term "behaviors" is being used in a broad sense here, to include thoughts, actions, emotions, and personality traits. Results of psychological testing are usually presented as scores and in a way that different persons, or the same person at various times, can be compared.

Several factors must be considered before a test can be used with a population. First, is it reliable, or in other words, is the test relatively free from error? Will an individual achieve a similar score on two separate testing occasions (assuming no learning or other change took place in the interim period between testings, including due to the experience of the testing itself)? While there are several sources of error in a score, the biggest is probably the error associated with poorly conceptualized or worded items. Whether or not the test is valid for the population and question at hand is the next matter to consider. A test may be valid for one use and be invalid for another. For example, we wouldn't use an intelligence test to gauge citizenship. Validity is not an all-or-none proposition. A test may be valid to an extent in one situation and less valid in another situation. In order for a test to be considered valid, its scales must not only behave well internally, but the test must actually differentiate among participants of known status.

There are many fine commercial-off-the-shelf (COTS) tests available for clinical use, but often their use is inadvisable due to a lack of appropriate norms for the aviator community (Retzlaff, King, McGlohn, and Callister, 1996). Gathering population-specific norms, on the other hand, is costly and time consuming. Practitioners are cautioned to use COTS psychological testing with care and to avoid over diagnosing psychopathology by comparing aviators to individuals in the general population (King, 1994), or missing cognitive impairment.

Case Vignette

A pilot in his early 30's is referred to an Otolaryngologist on account of task saturation and hearing loss. He is having difficulty staying ahead of his jet aircraft while flying and is missing radio calls. He had no difficulty while assigned to a trainer aircraft despite having approximately the same degree of hearing loss. He passes an "inflight hearing evaluation," but is noted to have difficulty under complex, demanding flying conditions. He is referred to a psychologist to assess his motivation to fly and to rule out a learning disability or other neuropsychological condition. The Halstead-Reitan Neuropsychological Battery reveals no central neuropsychological findings. His performance on the Speech-sounds Perception Test places him in the mildly impaired range. His Minnesota Multiphasic Personality Inventory-2 (MMPI-2) profile and other personality testing results are entirely within normal limits. His responses to diagnostic interviewing were likewise not suggestive of a DSM-III-R diagnosis. His motivation to fly, relatively late in its origin, is nonetheless strong and unconflicted. However, on the WAIS-R he achieves only average scores (when compared to the scores of the population at large). Intersubtest scatter was not significant; his abilities were roughly evenly developed. It seems likely that the trainer cockpit was more forgiving of his peripheral hearing deficit compared with faster jet demands.

Assessment of Fatigue

The ALAPS measures trait-specific factors, as previously described. The Sustained Operations Assessment Profile (SOAP; Retzlaff, King, Marsh, and French, 1997) measures the state-specific aspects of fatigue.

Inadequate rest or working through the normal sleep period can result in acute and chronic fatigue, leading to human error in military and civilian aviation. As military forces and other "down-sized" organizations, to include the deregulated airline industry, are pressed to do more with less, they may increasingly be faced with fatigue, leading to degraded performance and vulnerability to stress-related symptoms. Extended on-duty demands are referred to as Sustained Operations (SusOps), during which a long period on duty is required with, typically, no sleep breaks. An example might be ferrying fighter aircraft overseas, often requiring a 12-16 hour flight in a single seat fighter where sleep is not possible. Managing fatigue carefully is often vital to the success of these extended duty periods. As night vision goggles and other affordances turn night into day in the military realm, and as consumers become increasingly accustomed to 24-hours-per-day service in the civilian sector, the importance of understanding fatigue will grow. Individuals who must fly at night must not only not be expected to work a regular day shift, they must also not be reinforced for working during the day. These night fliers should be essentially banned from the office during day hours. Such an action requires that others (administrative support) likewise adjust their duty hours.

Needing to function during the nadir (low point) is another difficulty faced by all night-shift workers. For most tasks, the nadir falls between 0300 and 0600 hours of the circadian rhythm cycle (Klein, Wegmann, Athanassenas, Hohlweck, and Kuklinski, 1976). A problem frequently encountered in night fliers is their insistence to not alter their families' lifestyle, who remain on a typical schedule. Additionally, these aviators are likely to attempt to rapidly revert to a more conventional schedule during times when they're not night flying. Hence, they are likely to be in a continuos state of desynchrony. Violating the circadian rhythm seriously degrades performance, even when the skill is normally automatic and well established. The ability to respond to changing conditions is even further impaired. Microsleeps, or involuntary naps of two to three seconds in duration, are especially dangerous at the speeds of today's aircraft and other complex systems. Moreover, microsleeps are not restful and the individual does not even recognize that they have been sleeping; an EEG proves otherwise.

While brief periods of fatigue and poor sleep hygiene result in decrements in cognition and affect, some missions are always "up tempo." NASA shuttle missions, for example, combine one to two weeks of high operational tempo with acutely fatiguing, novel situations. The typical astronaut trains for two years for a mission. This training is scheduled down to the hour and becomes most onerous in the months prior to a launch. The excitement, combined with last minute changes and launch postponements, can be very fatiguing.

Traditional means to study fatigue have been objective performance tests and standardized mood surveys such as the Stanford Sleepiness Scale (SSS; Hoddes, Zarcone, Smythe, Phillips, & Dement, 1975), the Profile of Mood Survey (POMS; McNair, Lorr, & Droppelman, 1971) and Visual Analog Scales (VAS; Folstein & Luria, 1973). These metrics have distinct advantages and disadvantages. Objective performance tests, while often extremely fatigue sensitive, require stable performance levels to be achieved so that deviations from normal can be discerned (Schnieder, 1985). This process requires hours of practice; such demands of crew time are frequently difficult to accommodate. The SSS is extremely convenient but consists of only a single question with single response and fails to tap into other mood dimensions. The use of single items also results in limited reliability from a psychometric perspective. The POMS reliably measures six mood dimensions (fatigue, vigor, confusion, anger, tension, depression) but consists of 65 adjective evaluations. These measures require considerable time to administer and to score. VAS are frequently shorter in time requirement and allows the respondent to place a mark anywhere on a line between two extremes to indicate a response. While this method allows for a wider range of scores than either of the other mood measures, the unguided nature of the responses are open to considerable differences in interpretation. The resolution of this scale may also be overly artificial. The SSS and the VAS are typically single responses for each mood. The POMS has several questions to assess each mood and may thus strengthen the reliability of the mood assessment.

Folks cannot become "used" to sleep deprivation. A continuous state of acute fatigue leads to chronic fatigue, which is less well understood.

SOAP

SOAP is a 90-item test with a Likert scale of 1 to 5 (1- not at all, 5 - very much). Participants are asked to rate each of the items as they apply to them for "about the last hour or so." The figure below lists all test questions, which are presented to participants printed on the front and back of a legal-size form. As such, it is very compact and appears "short" to the test taker.

Each of the 10 scales have nine items which represent related aspects of the dimension. The items are presented as blocks under each of the scale names. This format differs from the typical psychological test, which usually randomizes items across scales and does not make explicit the scale names. The test usually only requires three to five minutes to complete. It is therefore well suited to situations requiring repeated administrations.

SOAP Scales

The scales of the test include three cognitive dimensions (*Poor Concentration, Boredom*, and *Slowed Reactions*), three affective dimensions (*Anxiety, Depression*, and *Irritability*), and four arousal dimensions (*Fatigue/ Low Energy, Poor Sleep, Work Frustration*, and *Physical Discomfort*).

Cognitive Dimensions
- *Poor Concentration*: High scorers report difficulty concentrating and paying attention. They have difficulty engaging in concentrated effort and consequently work inefficiently. They must repeat work and pay extra attention to complete tasks.

- *Boredom*: High scores are disinterested and complacent. They view the work as tedious and tiring. They no longer see "fun" in the workload and time appears to have slowed for them.

- *Slowed Reactions*: High scorers are cognitively and physically slowed. Additional effort is required to keep up with the task load. Things around them seem slowed and they are also slowed. Motor activity is impaired by the cognitive inputs and outputs.

Affective Dimensions
- *Anxiety*: High scorers feel nervous, anxious, and worried. Physically they are tense and jittery. Autonomically, they are vigilant and upset.

- *Depression*: High scorers feel depressed, unhappy, and sad. They are discouraged and lacking in enjoyment. They are concerned about their feelings and beginning to feel helpless.

- *Irritability*: High scorers are annoyed with others and feeling unfriendly. This includes the desire to be alone and away from others. They are impatient, disagreeable, and may be angry.

Arousal Dimensions

- *Fatigue / Low Energy*: High scorers are tired to the point of feeling burned out and worn out. They lack energy and doubt their ability to go on or perhaps even move.

- *Poor Sleep*: High scorers report sleep that was too little and of poor quality, leading to feeling sleepy and desiring sleep. Their eyes are closing and they may be nodding off. To compensate, they may be focusing on keeping their eyes open and bouncing or shaking.

- *Work Frustration*: High scorers are tired of working. They wish they didn't have to complete tasks and hope nothing new comes up. The work feels like a grind and drudgery. They may feel they need help to complete the job.

- *Physical Discomfort*: High scorers have sore muscles and are stiff. They want to get up and stretch. They may feel physically uncomfortable due to perspiration and binding clothing/gear. They feel the discomfort in their arms and legs, head, and eyes.

The following form, representing the SOAP, may be reproduced and used as a method of assessing fatigue. For best results, format on two sides of legal-size (8.5 inch x 14 inch) or A4 paper.

Instructions: Please rate each of the following as they apply to you **for about the last hour or so**, 1 means not at all, while 5 means very much.

1. *POOR CONCENTRATION*
1. DIFFICULTY CONCENTRATING
2. HARD TIME PAYING ATTENTION
3. CAN'T STAY AT A TASK
4. DISTRACTIBLE WHILE DOING THINGS
5. HAVING TO RE-READ MATERIAL
6. LOSING TRACK OF CONVERSATIONS
7. HAVING TO PAY EXTRA ATTENTION TO UNDERSTAND
8. DAYDREAMING
9. NOT THINKING STRAIGHT/ EFFICIENTLY

2. *BOREDOM*
1. NOTHING SEEMS INTERESTING
2. NOT CARING ABOUT WHAT IS HAPPENING
3. NONE OF THIS IS FUN ANYMORE
4. NOT INTERESTED IN WHAT IS HAPPENING
5. INDIFFERENT
6. BORED WITH MISSION
7. TIRED OF SAME OLD THING
8. THINGS ARE TEDIOUS
9. TIME IS PASSING TOO SLOWLY

3. *SLOWED REACTIONS*
1. NOT MOVING VERY MUCH
2. JUST LOOKING AROUND
3. HAVING TO THINK BEFORE ACTING
4. THINGS SEEM IN SLOW MOTION
5. ARMS FEEL HEAVY
6. REACTIONS ARE SLOWED
7. MOVEMENTS SEEM DELAYED
8. CAN'T KEEP UP WITH TASKS
9. RESPONDING TAKES EFFORT

4. *ANXIETY*
1. FEEL ANXIOUS
2. FEEL TENSE IN MUSCLES
3. FEEL NERVOUS
4. WORRIED ABOUT THINGS
5. MUSCLES ARE JITTERY
6. TAPPING FINGERS OR FOOT
7. STOMACH FEELS UPSET
8. JUMPY
9. VIGILANT

5. *DEPRESSION*
1. FEEL DEPRESSED
2. FEEL UNHAPPY
3. FEEL SAD
4. NOT ENJOYING THIS
5. FEEL DISCOURAGED
6. WISH I FELT HAPPIER
7. WORRIED I MIGHT NEVER FEEL BETTER
8. NO CONTROL OVER ANY OF THIS
9. WHAT I DO DOESN'T MATTER

6. *IRRITABILITY*
1. FEELING IRRITABLE
2. GENERALLY ANNOYED WITH OTHERS
3. PRETTY UNFRIENDLY/ TESTY
4. IMPATIENT
5. WOULD REALLY LIKE TO BE ALONE FOR A WHILE
6. GETTING ANGRY
7. DISAGREEABLE
8. WISH I WEREN'T WITH THESE PEOPLE
9. FEELING GROUCHY WITH PEOPLE

7. *FATIGUE/ LOW ENERGY*
1. VERY TIRED
2. REALLY FATIGUED
3. BURNED OUT
4. WORN OUT
5. FEELING EXHAUSTED
6. CAN'T GO ON MUCH LONGER
7. NOT ENERGETIC

8. TOO TIRED TO MOVE
9. FEEL DRAINED

8. *POOR SLEEP*
1. SLEEPY
2. WISH I HAD SLEPT LONGER
3. WISH I HAD SLEPT MORE SOUNDLY
4. COULD FALL ASLEEP RIGHT HERE
5. NODDING OFF
6. EYES ARE CLOSING
7. FORCING SELF TO KEEP EYES OPEN
8. BOUNCING, TAPPING, SHAKING TO STAY AWAKE
9. YAWNING

9. *WORK FRUSTRATION*
1. PREFER TO NOT WORK NOW
2. WISH I DIDN'T HAVE TO DO THINGS RIGHT NOW
3. HOPE NOTHING ELSE NEEDS TO BE DONE
4. TOO MUCH IS EXPECTED OF ME RIGHT NOW
5. TIRED OF WORKING
6. WORK FEELS LIKE DRUDGERY
7. FEELS LIKE A GRIND
8. WISH THE WORK WOULD END
9. WOULD LIKE HELP WITH THE WORK

10. *PHYSICAL DISCOMFORT*
1. MUSCLES HURT
2. BODY STIFF
3. CAN'T GET COMFORTABLE
4. ARMS AND LEGS HURT
5. WANT TO STRETCH
6. SKIN STICKY/ DIRTY
7. WHAT I'M WEARING IS UNCOMFORTABLE
8. HEAD ACHES
9. EYES TIRED

5 Providing Feedback to Pilots and Referral Sources (often Flight Surgeons or Commanders)

Being a Consultant

Generally speaking, the best avenue for success is for fledging aerospace clinical psychologists to serve as consultants to the typically better-established flight surgeon.

Pilots and other aviators welcome feedback, particularly when their wings and consequently their careers are on the line. Be sure to allow enough time to describe how scales were developed and validated; you will be impressed by their grasp of mathematical concepts. You will be impressed by the amount of information referred aviators request at the end of their evaluation, regardless of your recommendations.

Establishing an Operational Aeromedical Psychology Program (Being a Consultant to Flight Surgeons)

The best way to have a successful aerospace clinical psychology practice is to be a consultant to the flight surgeon's office or to an aeromedical examiner. Do NOT accept direct consultations from Wing commanders when functioning in flight psychologist mode. Also, only a flight surgeon has the authority to ground an aviator. You may need to remind people of this fact at strategic times, such as when attempting to establish rapport with a leery aviator.

Encourage and enable the flight surgeons to refer cases to you. Educate them about appropriate referrals to make. Mention that you are willing to tackle marital therapy cases (if you are, indeed, willing to do so and have the expertise). Aviators may have difficulty maintaining intimate relationships due to their controlling natures. A way to hook aviators into therapy is to initially see only the spouse and send messages home that will motivate the aviator to come in to tell his (her) side of the story. Remember, there are always three sides to every story: His, hers, and the truth (Krystyna Osinski-King, personal communication, Oct 19, 1989)!

Train the flight surgeons with whom you work to see a red flag when they hear about sleep problems. Dr. David R. Jones (personal communication, Nov 7, 1989) states that a sleep problem is the number one symptom for a troubled aviator.

Develop aviator-only smoking (nicotine) cessation sessions that do not use ground medications so that flying status is maintained. Conduct these programs away from the hospital; at the aviators' workplace, if possible. See Chapter 9.

Demonstrate your assessment abilities by offering to see individuals being considered for special duties. Remember, flight surgeons and other aeromedical physicians are typically involved with personnel standards decisions. Psychologists obviously have some unique skills to offer in the area of assessment.

Be discrete when you document in any aviator's medical records as seemingly innocuous observations may result in catasphopic career consequences. Learn to make good use of "V-codes" in the Diagnostic and Statistical Manual of Mental Disorders (American Psychiatric Association, 1991) such as life circumstance problem, marital problem, phase of life problem, etc. Even airsickness can be conceptualized as a V-code: occupational problem.

Getting Out Into the "Field"

Aerospace clinical psychology is not necessarily office-based and should involve working with people in the real-life workplace. You will need to spend as much time as possible at the wing. The more you are seen, the more you become a fixture and the less of a threat you will be. Make yourself a regular visitor to the Flight Safety Office. The "client" is not the person you interact with, rather, the client is the mission of the employer. Aerospace clinical psychology then is concerned primarily with mission accomplishment. Obviously, such an approach goes directly against the traditional role of a "provider" or clinician and is therefore likely to be misunderstood and possibly even resented by peers. It will be assumed that you are coasting during the time you are not in the mental health clinic. On an individual basis, you need to decide what price you are willing to pay to fulfill the role of flight psychologist. Accept that you may not be the most popular person in the mental health clinic. It is possible that you will be seen as a slack off who does not care about sick people. You may arouse jealousy. Deciding to under take this specialty may not enhance your career or your relationships with colleagues. It is therefore vital for you to work with other aviation-oriented psychologists even if they are hundreds of miles away.

Making Contacts/"Networking"

No task is more professionally demanding of your creativeness and ingenuity than that of flight psychologist. As a first contact, you may want to ally yourself with an aerospace physiologist, if available. Try to participate in any special training offered, first as a student and then as an instructor as appropriate. Try to get invited to as many meeting as possible, particularly those where "human factors" are discussed. Examples include any special monitoring status meetings about students. Consider doing briefing at squadron wives meeting, but before you open your trap, be sure you KNOW YOUR AUDIENCE (see "Speaking the Tongue"). Be prepared for a lot of joking. You may arouse anxiety in this high-functioning but relatively non-insightful crowd. Remember that it takes a good deal of comfort and familiarity for them to be able to make fun of you to your face.

Institutionalizing the Program

Do whatever you can to formalize your program. Prepare Operating Instructions, Letters of Agreement, and whatever else to legitimize or institutionalize your program. Pursue and accept any and all training opportunities, especially those that will increase your familiarization with the flying environment. Regularly visit air traffic controllers' duty station and talk to maintenance personnel. Consider taking flying lessons if you are genuinely interested in flying and have the time and funds to spare for this intensive undertaking.

Other tips:

- Develop some simple handouts and briefings and have them ready on short notice.

- Be professional when making presentations; use audio visual aids.

- Try to get some coverage from the local or base media (as appropriate) but remember to be humble and give credit where credit is due (flight surgeon, Chief of Hospital Services, Hospital Commander, Wing Commander, etc.). Submit articles for publication about your   program. Be sure to co-ordinate with your organization's public affairs function as appropriate. Don't allow yourself to be misquoted.

Remember, your population is not just aviators. Serve family members. A military career, or even a nonmilitary flying career, can challenge a marriage in many ways. When one or both of the spouses are aviators, the challenges multiply. After our study of female and male pilots (King, McGlohn, & Retzlaff, 1997), our only regret was not studying the unique subset of military aviators married to each other. Frequent travel (known as TDY - Temporary Duty), sudden deployments, missed family events, and the perception of forfeited financial opportunities contribute to family stress.

6 Providing Support: Critical Incident Stress Debriefing

Crisis Intervention

Butcher, as early as 1980, advocated immediate crisis intervention services following a cataclysmic aviation event. The typical reaction to such an atypical event follows a predictable course, to include:

- shock
- disorganization
- eventual reorganization

Reactions, however, may be delayed by hours or even weeks. The optimal services for affected individuals differ from traditional mental health care and include aiding victims and rescuers with necessities for daily living (food, shelter, and childcare arrangements). Additionally, relationships may be less formalized. For example, depending upon the premorbid personality structure of the targeted population, therapeutic touch may be more appropriate than in the typical client-therapist relationship.

Jones (1985) directed our attention to the plight of rescue/recovery workers, individuals he termed "secondary victims," after surveying the 592 USAF personnel involved with the aftermath of the mass suicide of almost 1,000 persons in Jonestown, Guyana. He found those exposed to the most bodies suffered the most dysphoria. He recommended older and experienced workers be used whether possible. If younger workers are used, they should be paired with older mentors whenever possible. Jones advocated for mental health workers to monitor the reactions of rescuers and to provide nonthreatening group opportunities for emotional support. Experience has shown it is best for the mental health professionals to remain as removed from the evidence of the disaster as possible, to include not viewing bodies, lest they become burdened with their own reactions.

The American Psychological Association (APA), in cooperation with the American Red Cross, developed the Disaster Response Network (DRN) to combat post traumatic stress disorder arising from disasters and crises. To date, over 2,000 psychologists have agreed to voluntarily provide onsite services to both disaster survivors and the relief workers who have assisted them. These volunteers provide services for those experiencing transitory stress and continuously assess affected individuals for the need for additional, referred services.

Interventions

- Pre-Exposure Preparation Training: Used when exposure to potentially traumatic events is expected; emphasizes the normalcy of feeling stress in abnormal situations. This training is not a remedial program to aid persons unable to handle stressful situations.

- Defusing: Putting a traumatic event into perspective by talking about it. If the emotional intensity is not defused, post traumatic stress disorder could result; defusings are primarily educational in nature.

- Debriefing: A group experience to address cognitive, emotional, and physical responses from exposure to a potentially traumatic event. Not initiated until 24 - 72 hours after the termination of the event (use defusing in the interim).

According to the APA, normal response to the abnormal situation of a sudden disaster include:

- Recurring thoughts or nightmares about the event
- Difficulty sleeping
- Changes in appetite
- Anxiety and fear, particularly when exposed to events or situations reminiscent of the trauma
- Memory difficulties, including difficulty remembering details of the trauma
- Unable to face certain aspects of the trauma; avoiding activities, places, persons reminiscent of the event
- Hypervigilant (Overly alert)
- Low energy

- Difficulties making decisions
- Spontaneously crying
- Feeling

  scattered / unable to focus on work and daily activities
  irritable, angry, resentful, and easily agitated
  depressed, sad
  emotionally numb, withdrawn from others
  extremely protective of, or fearful for, the safety of loved ones

While all the above are normal reactions to the abnormal situation of a traumatic event, it is also critical to respect the defense mechanism of denial, particularly the denial of rescue workers. Dissimilar to traditional mental health services, helping those acutely traumatized does not involve tearing down defenses. The best way to help is to provide a responsive support system that follows the lead of the traumatized person.

As difficult as it may be for rescue workers, ensure that they leave when their shift is over. Emphasize that they need to recharge their batteries to avoid completely burning out. Be sure they obtain as much rest as possible, eat well (nutritiously), exercise, and avoid misuse of substances (caffeine, nicotine, and alcohol). Rescue workers are of little use to others if they are hungry, tired, or otherwise not emotionally available. Breaks are necessary for anybody and lead to increased efficiency and health of the rescue worker.

Returning Home

Rescue workers must also be prepared to return to their normal routine upon termination of their involvement with the disaster. Despite your family and workplace protests that they need you to do certain things right away, get as much rest as you can before plunging back into the work that accumulated during your absence. Be patient with your family when they note your absence during significant family events (Junior's third birthday, Sally's loss of her first tooth) or domestic "crises" (car broke down, family cat died). Don't minimize their experiences as the cost of emphasizing the significance of your work- they are trying to convey that they missed you. Keep reunion expectations reasonable; life rarely matches our desires or Hollywood in emotional intensity. Don't be surprised if you find that you miss the excitement of the emergency work, wishing you could return (especially if it ongoing). Recall that those in your daily life also need you. Your children may have had an especially hard time understanding your absence. When I was at a mishap investigation thirty days after telling my family I would be gone for no more than three to seven days, my son had a hard time believing I would return in a timely fashion from any subsequent trips. Keep in touch with family members during your absence, but keep it light. Be careful not to frighten children with overly graphic details or stories, either during your absence or upon your return. Involve your family in preparations for disaster deployment to help them feel important and empowered. Be sure to seek help if you find that you're having difficulty readjusting after your deployment. Failure to help yourself will result in diminished capacity to help others in the future. Above all else, be sure to grieve and to talk with supportive friends, who may not be in your regular social circle.

On-scene Supervisors

When the rescue operation is over, be sure to have a formal termination ceremony as "Recognition by valued authority is a powerful antidote for perceived suffering" (Jones, 1985, *p*. 307). A final chance to talk and referral to potential sources for support, geographically convenient to the soon-to-be released rescue worker, are vital.

Resources

- American Psychological Association (APA): The Practice Directorate, 1-202-336-5800 can direct you to the various state psychological associations, who along with city and county psychological associations, can refer you to disaster trained psychologists in a given geographical region. For example, the New York State Psychological Association Disaster/Crisis Response Network maintains a strong program.

- Red Cross/Red Crescent: Check local phone book listings for your local chapter.

- International Critical Incident Stress Foundation: This private, nonprofit organization maintains a database of some 400 fully trained teams throughout the world. Phone 1-410-750-9600 24 hours per day for assistance.

- <u>National Organization for Victims Assistance:</u> 1-800-TRY-NOVA (1-800-6682; in Washington, DC metro area: 232-6682).

7 Teaching

You've Been Asked to Teach: Now What?

I've tried to make teaching easy for even the semi-motivated reader by providing some of the more popular topics in lesson plan format. Use the material provided in this book to develop rapport with flying squadrons. Pay careful attention to "Speaking the Tongue" (*p.* 52); it will pay rich dividends when you speak to aviators. Learn the needs of your local aviators and then develop lesson plans to cover what they need to know in the minimum amount of time possible. Above all else, prepare carefully for even the most seemingly casual of teaching opportunities. This group demands excellence and will readily detect any lack of preparation. If you're too busy to adequately prepare, then don't attempt to teach!

Know Your Audience

Above all else: Know your audience! Very early on in my aviation psychology career, I made the mistake of having a very frank group discussion with the wives of the student pilots and even some "significant others" of the international student pilots. They spontaneously engaged me in a wide variety of topics, including why they were "required" by the older wives to get up at any hour to prepare breakfast for their men, why they needed their husbands' consent to obtain birth control, and why the military was not more supportive of *their* careers. I suppose I did not give very diplomatic answers, as the next day my boss (the hospital commander) told me that the flying wing commander called and asked: "What's wrong with that damned psychologist?" My commander, to whom I remain grateful, advised me to speak with the flying wing commander's wife. Realizing that one does not approach a senior officer's wife in such a fashion, I instead elected to make an appointment with the wing commander himself. This meeting evolved into a three hour discussion and resulted in many appropriate referrals, via the flight surgeon's office. At the end of our discussion, I requested permission to speak with his wife to resolve any misunderstandings. His response: "I'll take care of her."

Better Communication Through Lesson Planning

Do not underestimate the value of a lesson plan, either as the presenter or the requester of a presentation. Below is a 12-step method to prepare a lesson plan. Such an exercise requires you to carefully consider your goals. Once we tried to extract such goals out of a requester who desired a repeating presentation to some occupational health residents. He refused, however, to co-operate in the development of a lesson plan. Tactfully (well, maybe not too tactfully) we tried to encourage him to view a lesson plan as a necessary component for a successful presentation and resisted his pressure to help him identify a presenter. Nevertheless, he found a presenter, but soon was complaining due to the rotating residents not hearing the same presentation.

Twelve-Step Lesson Plan Worksheet

1.)Lesson Title: _____

2.) Target Population: _____

3.) Learning Objective(s) {Evaluated via Samples of Behavior}:

_____

_____

_____

Teaching Guide

4.) Introduce Self :_____

5.) Attention-Getting Step: _____

_____

_____

6.) Motivation: _____

_____

7.) Overview: _____

_____

8.) Main Body of Presentation

9.) Conclusion :_____

_____

10.) Summary: _____

_____

11.) Remotivation: _____

_____

12.) Closure :_____

Lesson Plan Instructions

1.)   Self-explanatory.

2.)   Who will be the recipients of this information? What are their interests and capabilities?

3.)   What you want audience members to take away with them expressed in measurable or observable terms. Usually framed as "Identify" or "Perform." The formula is as follows: a.) Given condition b.) desired student behavior Example : (a) Provided a list of alternatives/ (b) student will identify the correct alternative.

4.)   Self-explanatory.

5.)   Joke, Story, Stunt, or Quote relevant to the lesson.

6.)   Brief audience as to how/why this lecture information is of
            interest/importance to them. Hypothetical example: "Our
            surveying suggests that new flight surgeons are sometimes
            challenged with the need to make an adequate referral for an
            aviator with a special needs child."

7.)   What you are about to say.

8.)   The time to say it!
            This section is always prepared in outline form. Audiovisual
            aids and planned overhead questions are specifically
            indicated. Also, indicate when to distribute any handouts.

                        Example:
                        I. Stress (Topic)
                          1. Autonomic Over-arousal (Major Point)
                            a. Airsickness (Subpoint)
                            b. Heart palpitations
                          2. Loss of situational awareness (Next major point)

                        Overhead Question: "Does anyone care to guess the
            result of loss of situational awareness while in a busy traffic
            pattern?"

                        II. Topic (Etc.)

            The above material would be prepared on the left side of the
            page. Individual variations (examples, personal experiences,
            etc.) are listed on the right side of the page. The content
            universe for testing purposes (as defined by the objective)
            are always drawn from the left column.

9.)   Any final new information is presented here Also the time to
            begin to wrap up and prepare your audience for their next
            lesson.

10.)   Reiterate what you've presented during the lesson.   Rehash all major points.

11.) Remind audience of importance of information for their purposes.

12.) Time to let audience know that the lesson is now complete. End with a joke (IF APPROPRIATE TO THE TOPIC!), give phone number again.

Audio-visual hardware required (Overhead projector, slide projector, projection monitor, etc.):

Audio-visual Software to be Used (Transparencies, slides, etc.):

Multiple choice questions with one correct alternative (indicated). These should flow from your objective(s).

Attach a copy of each handout to be utilized.

Often, you may be called upon to give a presentation on a one-time (or so it may initially seem) basis. Many presenters show up with either a pile of transparencies *or* a sheaf of papers. It may be most productive to combine the two approaches, eliminating the drawbacks of both and capitalizing upon the benefits of both. Tie the text into the slides; write a script rather than a manuscript. Bring two copies with you, one for your use and one for your hastily recruited, but eager-to-please, transparency handler. That way your every gesture or grimace or inflection will not be misinterpreted as a signal to change transparencies. An additional advantage is that you'll be able to read your slides without turning your back to your audience. While some may fear that a script robs one of spontaneity and perhaps even renders the presenter an automaton, such a strategy may actually foster greater spontaneity. If you have what you need to say directly in front of you, you are then free to read your audience and appropriately tailor your remarks, without fear of straying too far afield.

How To Talk To Pilots

A Light-Hearted Guide to "Speaking the Tongue"
Adapted from an unpublished article by
Col. (Dr.) Roger F. Landry, USAF, Retired

- Aircrew members are not stupid or lacking in adequate attention and concentration. They may appear rude and impulsive but in reality they are quite conforming. It's just that they shun abstractions and get bored easily. Ignore these realities at your peril!

- Be brief when you brief. Remember, as a human factors specialist, you represent a threat. The milk of aircrew kindness comes in very small containers and has an expiration date. If you are any good, they are likely to afford you only five or 10 minutes. If you are not particularly good, then you will have only a matter of seconds. If you insist on continuing, you will find yourself facing an increasingly hostile mob. An ugly scene may develop. Consider running for your life.

- Be funny. Aircrew like to have a good time and don't take themselves too seriously. Don't take yourself too seriously. Remember, seriousness is for shoe clerks. If aircrew members are not laughing or smiling at least once every ten minutes, they must be either grounded or dead (a distinction they don't make).

- Prepare so that you appear spontaneous. No, the preceding is not an oxymoron. Aircrew members love spontaneity even if it is contrived. Remember, these people are compulsive planners and leave little to chance.

- Know what you're taking about. BS artists are quickly spotted for what they are: TARGETS!

- Use aviation metaphors and colloquialisms freely if you know what you are talking about. Engage in malapropisms and you may thereafter find yourself ignored. If you don't how, then ask!

- Be prepared for some good-natured heckling. They wouldn't do it if they did not like you. The ultimate insult from these people is for them to ignore you. Remember, they're paying attention if they're engaging you.

- Watch any "wannabe" tendencies that you may be harboring. Remember, we are not pilots. The healthy motive for working in the aviation environment is to positively impact the flying mission and increase flight safety, not to give you an excuse to wear a flight suit.

8 Additional Assessment Tips

Interviewing

Aviators live by checklists and take comfort when they see others using them. Structured and semi-structured interviews, therefore, are useful when interviewing these folks. Determine the aviators', or would-be aviators', motivation to fly and history of that motivation. While motivation to fly can be usually traced to a very early age, sometimes aviators got the "flying bug" much later in life. Be aware, however, that clinical experience suggests that early motivation is usually more resilient. The structured interview suggested below is similar to the previously cited ARMA, without, however, any attempt at quantitative scoring.

Structured Interview to Assess Motivation to Fly:

When did you first become interested in flying?
Why do you fly/want to fly?
What fears do you have while thinking about flying or flying?
What do your friends and family think about your flying career?
Have you thought about the risks?
What are your goals for the future (see if they spontaneously mention activities involving flying, or not)?
What would you do if you if you couldn't fly?

General Interviewing Tips:

- An interview is not a conversation.
- Interviews are purposeful, directive, and non-reciprocal.
- Use limited self-revelation.
- Build rapport and the relationship.
- Listen.
- Restate and ask for clarification.

- Offer beverages and take breaks as needed.
- Ask open-ended questions.
- Encourage spontaneous recall.
- Start general, then get specific.
- Ask closed-ended questions as appropriate.
- Efficient for details, encourages recognition, but may be leading.
- Pay attention to the process as well as the content.
- How is this person relating to you? (Note tone of voice, posture, and facial expression.)
- Make a judgment about the level of cooperation you're getting.
- If they're not answering your questions, change your style.
- Be comfortable with pauses.
- Let him think, let him talk.
- Take your time and be complete.
- You don't want to have to go back and re-interview if you don't have to do so.
- Be aware of who the client is.
- Explain the uses of the information you are gathering.
- Do not promise absolute confidentiality if you cannot guarantee it.

Understanding the potential vulnerabilities of aviators can be a useful tool in rendering services to this extremely stressed population. Know the pressure points for this group and beware if they start to self-medicate, argue with spouse, or present you with physical problems that do not appear to have an organic basis. Such cases must, of course, be closely co-ordinated with medical colleagues, when being seen by a psychologist.

At some point, an aviator ceases to be an aviator and becomes a patient. While some clinicians may be tempted to keep obviously impaired aviators on flying status, David R. Jones makes the point that "flying isn't therapy" (personal communication, 7 Nov 91). Despite the possibility of a gloomy prognosis in cases involving numerous medical complaints, let us move on to consider aviator-specific interventions.

9   Interventions

First: What is "Stress" Anyway and Can (Should) Anything be Done About it?

As stated throughout this book, flying is an inherently stressful activity. Can this stress be eliminated?   It is neither possible nor desirable to eliminate all sources of stress. What is "stress" anyway?

General Adaptation Syndrome (Selye, 1978):

> Stress is a challenge to the organism
> 1.)   Alarm Reaction
> 2.)   Stage of Resistance
> 3.)   Stage of Exhaustion (Distress)

Stress is a signal for action. Some stress is necessary for optimal performance. Yerkes and Dodson (1908) described and inverted U curve with optimum performance occurring with moderate amounts of stress (or arousal). Why is it important to manage stress? Too much unprocessed stress and we become overwhelmed (distressed). Not enough "stress," however, and we also exhibit less than adequate performance. Distress can inhibit decision-making, impair performance, and lead to a breakdown in physical integrity (illness). As change occurs, positive or negative, the risk of being involved with a mishap or becoming physically ill increases (Haakonson, 1980; Brown and McGill, 1989). Why is this state of affairs the case? Success (positive changes) places new demands on us to "keep it up" or even outshine ourselves.

Stress Management

What are stress management techniques? The techniques are simple but require you to actually employ them.

- Practice good nutrition.

- Get adequate sleep and other "down time" (give yourself permission to do little or nothing sometimes).

- Exercise. Stay in good physical shape (make time in your day for some exercise, even if only a thirty-minute walk).

- Emphasize relationships.

- Get rid of the things that don't work in your life. Repair or discard broken appliances. Consider doing the same with dysfunctional relationships.

What are you already doing to manage stress? Remember, you don't have to control your emotions to manage stress.

Anger (and Alcohol) Management

You may consider anger dangerous - a sign of bad manners or even insanity. Anger, however, is different from aggression. Aggression is the way some people choose to express anger. The image of "The Incredible Hulk," with anger itself leading directly to total loss of control, has given anger a bad reputation.

You may allow others to push your anger buttons, but they do not MAKE you angry. It's important to realize WHEN others are pushing these buttons, and to rationally decide what you're going to do about it (maybe nothing). You have the RIGHT to be angry, but you don't have any right to violate others' rights. Anger has a strong physiological component, ignore anger at your peril (illness: impotence, high blood pressure, and gastrointestinal problems). Perhaps you've heard of getting "pissed off." The "type A personalities" who die early expire not from their drive to succeed; rather their killer is unverbalized anger (and impatience).

Denying anger consumes energy and may come out in other, unintended, ways. Denied anger may erupt unexpectedly and be directed at another (unintended) audience. Some people are so "successful" at suppressing anger they become numb to all emotions, even positive ones. Anger that is not appropriately managed may lead to alcohol abuse, the consequences of which may further increase angry emotions.

If you think the typical aviator is evasive and defensive, wait until you run up against an alcoholic. If you are not skilled in the diagnosis and treatment of alcohol and other addictive disorders, be sure to seek immediate help. An useful screening tool is the CAGE Questionnaire (Cutting down, Annoyance by criticism, Guilty feelings, and Eye-openers; Ewing, 1984). Aviators are not immune from addictive disorders and it's well known that work performance is typically the last area of functioning to be impaired by an addiction. Beware of the common statement that someone is not alcoholic because they "Don't drink anymore than me." First, the important factor is what are the consequences when someone drinks, not the amount and secondly, perhaps the person making that statement is, himself, alcoholic. The organization Alcoholics Anonymous (AA) is a tremendous aid in an alcoholic's recovery process, when that individual is ready to change, and meetings are available around the world. Furthermore, participation is truly anonymous, as its name implies. "Birds of a Feather" is a group designed expressly for aviators. Many aviators may be conscientiously drinking their prescribed glass or two of wine per day in an effort to forestall cardiac problems, when in reality they may not be part of the group who can benefit from such a regiment. Despite their good intentions, some members of this group may develop alcoholism due to a genetic predisposition. Furthermore, "bottle to throttle" rules (typically 12 hours) often ignore the residual effects of heavy alcohol use *before* the 12th hour. Given a choice, it may be better to fly with a crew that all just had a bottle of beer than with a crew seriously hung over!

Anger is different from aggression and "madness." Aggression is the way some people choose to express anger. The image of "The Incredible Hulk," with anger itself leading directly to total loss of control, has given anger a bad reputation.

You may allow others to push your anger buttons, but they do not MAKE you angry. It's important to realize WHEN others are pushing these buttons, and to rationally decide what you're going to do about it (maybe nothing). You have the RIGHT to be angry, but you don't have the right to violate others' rights.

Anger has a strong physiological component, ignore anger at your peril. The "type A personalities" who die early expire not from their drive to succeed, rather unverbalized anger (and impatience) is their killer. Denying anger consumes energy and may come out in other, unintended, ways. Denied anger may erupt unexpectedly and be directed at another audience.

Some people are so "successful" at suppressing anger they become numb to all emotions, even positive ones. They may become Mr. Spock (intellectualize) or may believe have no right to be angry (low self-esteem or over socialized).

The Cure?

- Recognize when you're angry.
- Learn to modulate your anger/Reclaim control.
- If necessary, seek out assertiveness/negotiation training.
- Pick your battles.
- Divest your ego (win-win). Go for solutions, not victories!

Other Fixes

- Train yourself to take several deep breaths BEFORE you respond.
- If you feel violent, take a time-out to consider your options.
- Avoid alcohol at those times (fuel for acting out).
- Don't guess or mind read what someone else is experiencing. Ask them!
- Learn to talk about your feelings by using "I statements" rather than "YOU accusations."
- "*I'm* angry," NOT "*You* make me mad."
- "*I* don't like that," NOT "*You* shouldn't do that."
- The APPROPRIATE expression of anger has not killed anyone (at least no one who didn't have it coming).
- Learn to deal with others' anger. Listening and reflecting back what you hear is more effective than shouting back insults.
- Acknowledge that sometimes you're powerless over situations and you'll paradoxically gain control over those situations.

Case Vignette

An experienced pilot is promoted to a supervisory job, thrusting much more responsibility upon him. His wife notices an increase in his consumption of alcohol. He becomes defensive when she mentions her concerns to him. The tension escalates with his wife until during one heated exchange, he nearly strikes her. Formerly a very cautious pilot, he violates minimum altitude requirements during training missions.

Several fellow pilots note a subtle change in his behavior. Some are so concerned that they mention it to the chief pilot, who requests to see him. During the course of their conversation he breaks his right hand on the chief pilot's desk.

What's up with this pilot and what should be done to assist him?

Case Vignette

An experienced commuter pilot with a 10-year history of drinking approximately three to four scotches per evening is evaluated for alcohol dependence. She relates a pattern of self-imposed rules, such as never drinking at all during the holiday season nor in the presence of her supervisor, and rigidly observes a "bottle to throttle" rule. She has received a drinking while intoxicated citation (but admits she drove at other times while possibly intoxicated), notes increased tolerance, and frequent blackouts. She finally hit her "rock bottom" after being unable to attend a social event with her husband. She self-identified to an aeromedical examiner and attended an inpatient 28-day treatment program, with her husband participating actively during the family week. Upon interview she maintains that she remains active in her rehabilitation, attending Alcoholics Anonymous (AA) four times per week. She verbalizes that she can never have even one drink without risking her recovery program. Her husband remains supportive and is actively pursuing his own recovery program via Al Anon.

What is this woman's prognosis? Would she be a good risk to be returned to flying status?

Sleep

Irritability, inattention, and loss of situational awareness can result from lack of sleep or even sleep deficits. Most adults need 7.5 to 8 hours or more of restful (undisturbed) sleep. Rapid Eye Movement (REM) sleep is the most restful sleep despite its seemingly restless quality. The bulk of REM sleep occurs during the last third of the sleep period.

Sleep Hygiene

- Get a steady daily amount of exercise, which deepens sleep. Do not widely vary bedtimes and especially rising times, even on weekends and other down periods.
- Do not exercise immediately before retiring.
- Keep bedroom temperature moderate; excessively warm rooms disturb sleep.
- Avoid chronic use of sleep preparations. The rebound effect can be severe.
- Eat a light snack before bedtime; hunger may disturb sleep.
- Avoid drinking excessive amounts immediately before retiring to minimize trips to the bathroom.
- Minimize or avoid caffeinated beverages, including soft drinks and foods with caffeine (chocolate).
- Minimize or avoid use of alcohol. Although alcohol may seem to help tense individuals relax, sleep tends to be shallow and restless.
- Always get up at a regular time, regardless of how little you've slept.
- Avoid tobacco, nicotine is a stimulant.
- Get up if you can't fall asleep; you can't force yourself to get to sleep. Individuals who get angry and frustrated because they cannot sleep should instead leave their bed and do something nonstimulating, such as read boring material or watch mindless television. Do not engage in any stimulating activity. Return to bed only when sleepy.
- Put your clock under the bed or otherwise obscure your view of it; don't continue to check it as you remain awake. Avoid calculating the hours (minutes) remaining before you must get up.
- Do not nap if you're experiencing insomnia. Napping will just make sleeping later more difficult.
- Separate sleeping and studying/reading areas. You will have difficulty falling asleep if you read or study in your bed (or you may have difficulty remaining alert while reading/studying).
- Consider ear plugs if you need to sleep at usual times due to shift work or if you rest in a noisy environment.
- Take a hot bath about two hours before retiring (sleep occurs and deepens as body temperature drops).
- When all else fails: Drink warm milk!

Holiday Blues

Is the Christmas holiday season the season to be jolly? Not necessarily! For many individuals and aviators aren't exempt, the period from Thanksgiving to New Years is the saddest and most stressful time of the year. The "holiday blues" may be particularly intense for military aviators due to family separations, limited flying hours (due to inclement weather and mandatory downtime), and perhaps limited financial resources.

While personal budgets are stretched to the breaking point and time is in short supply, painful memories of past holiday disappointments may be rekindled. All the while, we may think everyone else is having a joyous time. This perception can make those affected feel even more isolated and lonely.

To add fuel to this already combustible mix, alcohol often flows freely at this time. In fact, alcohol is often referred to as "cheer" at this time of year. Many an inexperienced drinker have turned to this seeming anesthetic and ended up feeling even worse, perhaps with the added stressor of increased family difficulties or even legal problems. While few may realize it, alcohol is actually a depressant and its ability to brighten our spirits is fleeting. Alcohol may seem to be a stimulant, but in reality it's a cortical inhibitor, as anyone who has ever placed a lampshade on their head can attest. Although the media may portray drinking as a glamorous holiday tradition, the results may end up being otherwise. In addition to increasing feelings of sadness, alcohol is often a factor in holiday suicides and traffic fatalities. Not very glamorous, is it?

Loneliness is a common holiday experience. While it is important to not withdraw from others, surrounding yourself with meaningless relationships with others won't cure loneliness. It is possible to be lonely in the middle of a crowd.

On the other hand, aviators may find themselves far from family and friends, maybe even lacking adequate social supports. These separations may be even more painful when others seem to be enjoying family gatherings.

The holidays often remind us of former times spent with loved ones who have since died. Divorced aviators, particularly those separated from their children, also face particularly difficult times.

So, how to cope with this gloomy picture? To cope effectively with the added pressures of the holiday season, set realistic expectations. Accept that you cannot be all things to all people. Schedule time for yourself. Physically exhausting yourself and spending beyond your means will only result in resentment and increased unhappiness.

Encourage friends to also set more realistic expectations. Often it's wise to agree not to exchange presents with even the closest of friends. Lower expectations of yourself and others. Look out for those who cannot be with their young children during the holidays.

Many folks expect the holidays to be a magical time. With expectations running so high, disappointments are inevitable. For example, the holidays cannot undo years of family tension. It is far better to accept situations for what they are rather than force conflictual family members to interact with each other. Forget what you see on television, real life isn't quite so simple.

While sadness during the holidays can be a temporary and normal feeling, and not a sign of true (clinical) depression, seek professional help if the sadness doesn't go away after the holidays or if it becomes unbearable. Above all else, let someone (friend, chaplain, flight surgeon, whomever) know if you're hurting and you'll live to fly another day.

Weight Management

Why is there a section on weight management in a book on aerospace clinical psychology? Is it just a blatant attempt to increase book sales, similar to the ploy used by tabloids ("Lose 100 pounds an hour following the ALIEN DIET!")? Actually, the motivation to include such information is due to the relatively strict weight standards both military and civilian aviators face. Aviators, as an occupational hazard, are also challenged to find nutritious meals at the times and places they need them.

Most aviators are far from obese, but may find that they need to lose some weight and, more importantly, need to keep it off. Before beginning a weight-reducing diet, it is important to be sure that the time is right and you have a reasonable chance of success. It is more difficult to stick to a diet if you begin a week before your birthday or before holidays associated with food. What holiday isn't associated with food?

Repeatedly failing at dieting can result in feeling badly about yourself, and emerging research suggests such dieting is counter-productive. Losing and gaining weight is called "yo-yo" dieting and results in your body adjusting itself so that future weight loss is more difficult. We are beginning to learn that the body wishes to stay at a fixed weight. This mechanism was very useful when famine was a real and persistent problem. Yo-yo dieting teaches the body to use available calories more efficiently. This efficiency is similar to a motorist driving 40-45 miles per hour on the highway to conserve fuel. Similarly, eating less than three meals per day or being on an extremely low calorie diet teaches the body to make the most of available calories. Skipping meals will not usually result in weight loss! A person who consistently skips meals not only usually eats more in the course of a day, but also teaches his body to utilize calories more efficiently.

Once you're sure the time is right to begin to lose weight, set a reasonable goal weight. It's better to initially set a goal that is too high than to set one that is too low. It can be lowered later if necessary. For example, if you weigh 210 pounds and you think you should weigh 145 pounds (and one of those insurance company charts roughly agrees with you), a reasonable goal might be 160 or 165 pounds. When you reach that goal you can then re-evaluate your goal. In fact it may be a good idea to set a series of goal weights. Remember, you are more likely to stick to a diet when you achieve small victories than when you throw yourself into a next-to-impossible task. When you do reach a goal, or even lose a single pound, be sure to give yourself a reward. Be careful, however, of giving yourself food rewards as that is obviously counterproductive. Instead, do something for yourself that you normally would not do. For example, see the movie you've wanted to see or begin to accumulate that new, smaller-sized, wardrobe you're going to need. If money is in short supply, then make the reward something that money can't buy: For example, a specified period of relaxation, a nice warm bath, or perhaps a visit to a friend.

Research has also shown that overweight people tend to eat according to the clock or depending on other external cues rather than when actually hungry. Overweight people are likely to eat when they see others eating, regardless of when they last ate. Likewise, overweight people tend to mislabel internal cues. For example, any discomfort felt when dieting tends to be associated with the possibility of their getting sick or fainting if they don't eat immediately

It is best to weigh yourself weekly, rather than daily or hourly. You will find it more tempting when dieting to weigh yourself frequently than to eat forbidden foods. The best way to get discouraged, however, is to notice a lack of weight loss from one day to the next. Body weight actually fluctuates during the course of a day and thus weighing yourself too regularly can be deceptive. It is a good idea to keep a written record of your weight. Make it a point to note your weekly weight regardless of whether you have lost weight or not. You can keep your record private or decide to display it openly. The choice is yours. Of course it is best to weigh yourself under constant conditions: at about the same time of the day, with about the same amount of clothes or nude, and on the same scale.

If you plan to lose anymore than a small amount of weight, plan to consult a physician. He or she can not only help you arrive at an ideal weight, not everyone should compare themselves to those weight charts, but can also assess your general health and investigate any underlying causes for obesity such as an under-active thyroid. Physicians can also guide those who should not lose weight too rapidly. Most important, the physician needs to reassure you that weight loss will be safe and you will not die in the process. Do not fall victim, however, to those physicians who offer magic pills. The careful observer usually notes that these physicians are sometimes overweight themselves!

A support group can be an important part of losing weight. It helps to recognize that others have faced and continue to face the obstacles that seem insurmountable. Learning how others cope with challenges can help us more effectively handle the challenges. Sometimes it is necessary to consult a psychotherapist for short-term psychotherapy to deal with ambivalence surrounding weight loss. Losing weight can result in dramatic changes in a life. For example, the formerly unpopular young lady may suddenly find herself to be the object of the interest of young men. Sometimes, however, losing weight fails to resolve problems that were blamed on being overweight. For example, merely losing weight will not automatically supply anyone with social skills. An overweight person may actually have to evaluate whether they are eating due to emotional strain, boredom, depression or anxiety.

Remember not to turn little "disasters" into big disasters. It is not the end of the world if you lapse. While it is better to stick closely to the diet, if you stray don't convince yourself that your entire diet is ruined for the day and you may as well "pig-out" for the rest of the day (or week or month). This strategy will be helpful after you reach your desired weight so that you can learn to deal with occasional treats. Unlike the alcoholic and his need to totally abstain from drinking, an overweight person cannot realistically and safely decide to never eat again!

Your best friend when dieting can actually be good ol' water. It is important to drink plenty of water. Not only will it give you a feeling of fullness, but it also will prevent dehydration. Just remember to avoid foods high in sodium (salt) or you will "hold" the water and you will lead yourself to believe that you're not losing weight.

It is also important to remember that it is best to lose weight gradually. Fad diets can have dramatic short-term results, but do not result in sensible eating habits and may even result in your becoming heavier than ever. The only way to lose weight is to consume fewer calories that you burn. Period. That is why exercise is an important part of a weight loss plan. Do not, however, overestimate exercise's ability to burn calories.

While it is important to avoid certain foods, particularly those with a high percentage of fats or simple carbohydrates (sugars and starches), it is also important to alter behavior around food. For example, don't eat while reading. People who simultaneously eat and read tend to overeat, as they do not fully realize how much they are eating. Drinking alcohol presents a double whammy as alcohol in any form represents "empty" calories and is usually associated with eating indiscriminately. It is harder to turn down cocktail peanuts after having a few cocktails. It is best to avoid having tempting snacks in the home. If others in the family, however, feel the need for these items, request that they not have them in a place easily accessible to you. It may be a good idea to keep these items securely wrapped up so that getting to them requires quite an effort. This effort gives you a chance to decide if you really want the snack and breaks the pattern of reflexively turning to snacks. By the way, a dieter should restrict all his eating to the appropriate dining area of the house. Do not eat in the recreation room in front of the television. Certainly do not eat standing in front of the open refrigerator! Meal times should be as stress-free and relaxed as possible. It may not be a good idea to discuss family problems at the dinner table, despite family traditions.

A reducing diet should actually be a basis for a lifetime of sensible, healthy eating. What makes a reducing diet a reducing diet is the limitation of fats (butter, oils, fried foods, fatty meats), empty calories (alcohol, refined sugar), and simple carbohydrates (heavily refined flour and other overly processed foods). Be very careful of foods that disguise themselves as "diet" foods but which are actually reduced portions of sugar-laden, nutritionally worthless foods. If you are uncertain about what a sensible diet consists of, consult a dietitian or a nutritionist.

The decision to lose weight is a brave one. Sometimes this decision is about half the battle. The period right after the beginning of a diet is the most difficult time. When a person starts dieting, it is likely that there will be a rather rapid weight loss. This weight is actually fluid loss due to the restriction of salt and possibly the consumption of diuretics such as tea. The next phase may actually see a slight increase in weight due to the body replacing some of the lost water. This phase is a critical point. Dieters frequently get discouraged at this point and grow bored of the routine of dieting that started as such a novelty. It is also important for dieters who are exercising vigorously to realize that any muscle they gain is denser than the fat they are losing, hence while they are not losing weight, inches will be lost in the desired, or rather the undesired, places and bulk will be added to the desired places.

When you reach your final goal weight, be sure to rid yourself of your "fat clothes." Donate them to a charity or the state hospital if you wish. If you're not opposed to having a little fun, consider a ceremonial burning. Although this procedure is less practical than donating the clothes to the needy, it is more fun and maybe even more therapeutic. After you achieve your goal, realize that your job is just beginning. You will be dealing with well-meaning people who encourage you to eat, citing the need to maintain your health or reminding you that you have lost your weight. Remember, the key to successful weight reduction is keeping the weight off.

The task before you is difficult, but by no means impossible. While it is true that people who have been overweight since childhood may have more adipose tissue and thus are more prone to remain overweight in adulthood, nobody has to remain a prisoner of flesh because of adipose tissue. The task is to keep the adipose tissue as vacant as possible, sort of similar to evicting the tenants of an apartment.

Nicotine Cessation

Nicotine cessation is of concern to aviators due to the diminishing opportunities to smoke, particularly in the aviation environment. Nicotine is particularly hazardous to those who fly at night as it reduces blood supply, and hence oxygenation, reducing visual acuity. Withdrawal from nicotine, however, presents its own problems for an aviator, such as irritability and reduction in cognitive functioning. Why do normally self-controlled folks become hooked on cigarettes or other delivery vehicles for nicotine?

Three Ways to be Addicted to Cigarettes

- Chemical addiction to nicotine - Nicotine becomes part of the user's physiology (nicotine gum or other slow taper methods may help with this form of addiction).

- Habit - Automatic behavioral ritual, such as having a cigarette with coffee or with a drink or after sex.

- Psychological addiction - using cigarettes to manage stress (but nicotine is actually a stimulant).

Two Ways to End Use of Nicotine-containing Products

- Gradual reduction in use of nicotine products (tapering off use).

- Cold turkey - Set a quit date and then just stop on that date.

To counter your habit patterns and possible ambivalence about quitting, be sure to enlist as much social support as possible (probably from your non-smoking buddies), get rid of your nicotine products, and try to avoid the situations where you're most vulnerable to back sliding (true for all addictions). Be sure to continue to take the breaks you used to call smoke breaks and continue your diagrammatic breathing - this time without cigarettes! For those in the US, call the American Cancer Society at 1-800-227-2345 for nicotine cessation programs.

If you're worried about gaining weight after quitting smoking, develop the positive addiction of exercise. Positive addictions differ from negative addictions, in that positive addictions are painful while we're doing them but we realize long-term benefits. Negative addictions are pleasurable while we engage in them, but result in long-term negative consequences.

10 Motivation and Fear

Manifestation of Apprehension

Manifestation of Apprehension (MOA) is a state of psychological anxiety, apprehension, and (or) physical impairment exhibited by students toward their training environment.

Symptoms

- Passive or active airsickness.
- Insomnia.
- Loss of appetite.
- Anxiety or tension related to the flying environment.
- Deterioration of performance or underachievement.
- May be a history of vague medical complaints, frequent flight surgeon visits, lack of preparation, daydreaming.

Identification

- Subjective evaluation of the instructor.

Indications to the Instructor Pilot

- Demotivation (decrease in motivation).
- Negative attitude toward learning.
- Loss of sense of humor.
- Personality changes.

In the USAF, if there are no psychological or physical problems, the student is deemed medically qualified for flying duties. If the condition does not improve, administrative, not medical, elimination occurs.

Fear of Flying (FOF)

- Non-phobic fear of flying is an administrative, not a medical, problem.
- *Phobic* fear of flying is an anxiety disorder, and thus medically disqualifying.

Phobia

- Person recognizes the fear as irrational.
- Nevertheless, attempts to avoid the situation.

"Flying phobia" is grounds for disqualification from continued military service if of sufficient magnitude to preclude transportation by military air transportation. FOF is dealt with administratively unless the proximate result of a psychotic disorder or a bona fide primary "neurotic disorder."

When a trained (rated) USAF aviator is diagnosed with Fear of Flying, he or she must appear before a Flying Evaluation Board (FEB). Banked pilots (trained pilots who are temporarily placed in nonflying jobs due to a lack of available cockpits) who are "coming out of the bank" (being reassigned to flying duties) are eligible for FEB.

It is important to explore the motivation to fly and the history of the motivation to fly. Maybe they were not very motivated to begin with. It's also important to recognize that motivation to fly changes over the course of a career and a lifetime. Normative for a middle-aged (defined as over 30 in a military population) aviator to have less motivation to fly due to increasing family responsibilities and possibly loss of one or more friends to aircraft accidents.

3 Questions (from Strongin, 1987)

- Do the symptoms stem from a preexisting disorder?
- What are the situational stressors?
- Have life circumstances temporarily altered motivation and defenses?

When seeing a patient with suspected MOA or FOF, ask what they will do if they don't go back to flying status, and then wait...

The defining characteristics of simple phobia are an irrational fear of the phobic stimulus and attempts to avoid the anxiety-provoking stimulus. Phobic reactions must be distinguished from fear, which is based on a more realistic appraisal. Empirically based interventions, such as systematic desensitization, have not historically been applied to fear of flying, despite their well-documented efficacy. This unfortunate disconnect may be due to the reluctance of otherwise adequately to high functioning individuals to seek the assistance of mental health providers. Conversely, these providers typically do not practice in the world of work, but instead tend to consult with patients who are more globally impaired.

Case Vignette

A young enlisted man is referred by his commander with the complaint of difficulty wearing a respirator while working in a hazardous environment. He complains of feelings of "claustrophobia" and states that he feels as though he can't breathe while wearing the mask-like apparatus. Although his commander states that he is motivated to overcome his condition, on interview he reports anxiety when even thinking about wearing the apparatus. Further interviewing does not reveal evidence to support claustrophobia, but does support a diagnosis of simple phobia, mask anxiety. He is offered treatment for his condition, but consistent with the diagnosis, declines to participate. His decision is supported by the clinician assessing him since treatment would require extreme motivation and an abortive treatment effort would result in a strengthening of his phobia condition (due to the reduction of anxiety and hence reinforcement of avoiding the phobic stimulus). The commander supports the disposition, although reluctantly when she learns that he is now unfit for continued military service due to his inability to wear a gas mask (and hence be worldwide qualified). He is consequently medically separated from military service.

11 Airsickness: Prevention and Management

The Nature of the Problem

Take any Joe or Jane Doe off the street. Hang 40 pounds on their back. Strap them into a device from the local torture shop. Then, encase their head in a tight plastic bucket and make them suck air through a hose. Now, stuff them in a small, hot decompression chamber that smells like kerosene. Jerk them around in all three directions at the same time. Flash a kaleidoscope of strange pictures all around them. Fill their ears with various beeps, hums, thumps, and numerous voices babbling rapidly in a foreign language. After all this, make them play Space Invader and convince them if they lose, they'll be burned to death!

Airsickness can be devastating to an aviator's career self-esteem. The very new aviator may end up finished before even really starting. The techniques to manage, or even prevent, airsickness, however, are simple. With a relatively small investment of time spent learning these techniques, you can help an aviator return to the sky. It is gratifying and cost-effective to return a formerly airsick aviator to the sky.

Origins

There are multiple causes for the response of airsickness. Each individual case is likely to be a combination of the following:

- Adaptation problem - We were not designed to fly. Being slammed around in three directions is unfamiliar to us. Acceleration, unfamiliar aircraft attitudes, and environmental stressors present in the aviation environment present adaptational challenges.

- Sympathetic Nervous System over-arousal - The body gears itself up to meet the challenges previously listed. A frequent problem is systemic overshoot resulting in rapid breathing, heart palpitations, sweating, etc. Basically the "Fight or Flight" response.

- Conflicting information - The brain receives conflicting information from the visual and vestibular systems during motion stimulation. This mismatched information is interpreted by the brain in much the same way that cues about poisoned food are   processed. The responses of general nausea and emesis are therefore not unreasonable.

- Anxiety regarding performance - Aviators, similar to any other high achievers, want to do very well. It is possible, however, to become over-motivated. The result, sadly, is a decrement in performance. Think about studying unusually hard for an exam and "choking" while taking it.

- Manifestation of Apprehension - Freud stated: "The symptom has a sense." Airsickness is sometimes an indication that a person would rather NOT be flying and may, in fact, even be afraid to fly. Rather then consciously experiencing the fear and recognizing their desire not to fly, a physical symptom is unconsciously manufactured that prevents flying. This cause is rare in student pilots, probably even rarer almost by definition) in experienced aviators.

- Low motivation - Airsick aviators are sometimes poorly motivated.   Some discomfort is inherent in the flying environment and even contributes to the fun of flying. When an individual consistently complains about the discomforts of flying, suspect poor motivation and consequent lack of tolerance to learning how to deal with the challenges of flight. Low motivation is much more common than manifestation of apprehension in the airsick student pilot population.

Types of Airsickness (Both result in a deviation in the mission profile.)

- Active - Characterized by rapid heart rate, sweating, excessive salivation, cold hands and feet (literally) and nausea finally culminating in, with any luck, filling the bag!

- Passive - All of the above symptoms except emesis (vomitting) does not occur. A deviation in the mission profile occurs due to the nausea and/or discomfort.

Helping Aviators Avoid or Manage Airsickness

Phase 1 (Prevent/Educate)

Start as early as possible by exposing all incoming students to prevention/ management techniques, before they fly! (Stress Talk, Anti-Airsickness Behavioral Checklist, A-B-C and sleep hygiene rules)

Talk to Aviators before they Commence Flying Training (Phase 1)

Airsickness is often a stress response that occurs when stressors are temporarily overwhelming one's ability to deal with them. Managing stress, both on the ground and in the air, is an effective method for avoiding or overcoming airsickness.

Stress cannot be avoided, but can be managed. Flying training has many stressors, those that are "negative" and those that are "positive." Your dollar (orientation) ride, although a positive event, will probably be stressful due to it being your introduction to an aircraft you've not yet mastered. While you will not be expected to fly during this ride, being out of control is itself quite stressful. Receiving your wings in a bout a year will also be stressful due to the change in status it involves, in addition the probably need to move and learn to fly yet another aircraft. The increased heart rate you will experience on the day/night you receive your wings is no different from a physiological standpoint from being involved with an in-flight emergency. Both events will result in increased heart rate, increased muscle tension, and sweating. Rather than thinking in terms of "bad" stress, let's think in terms of situational demands and life change units.

Some stress is necessary for optimal performance. Too much stress and we become over-whelmed. Not enough stress and what happens? (We become bored.) Excessive stress, however, can inhibit decision-making and impair pilot performance. It has been found by the Dutch airline KLM (Haakonson, 1980) that as the amount of stress increases, whether from positive or negative events, the risk of being involved with a mishap increases. Airsickness can also be a result of not managing stress; the stresses of flying and the stressors on the ground.

Stress and worry can inhibit decision-making and impair pilot judgment and performance. Inappropriate handling of stress can lead us to have tunnel vision - a narrow point of view or perception of options. Repeatedly, aviators involved with emergency situations talk of failing to appreciate all their options. Stress does not emerge solely from negative events. Any deviation can result in a stress response, particularly for individuals, such as yourselves, who value predictability. As the number of Life Change Units increase, the risk of being involved with an accident increases. One of the reasons for this finding may be due to your tendency, as a group, to act out rather than look within when confronted with distracters (family problems, financial concerns, conflicts with co-workers). While it is important to act in an aircraft emergency, you must first accurately assess the situation. The same is true in domestic emergencies. We all operate best at some moderate level of stress; below which we are bored and above which we are overwhelmed, maybe even panicked.

The methods to manage stress are actually quite simple, but require practice and consistency, similar to learning to interact with complex machinery. Stress management is all about getting proper rest, good nutrition, regular exercise, and even occasionally having fun. Although the training program on which you are about to embark will be demanding and will require your utmost attention, it will also be necessary to have some fun this year and sometimes even allow yourself to do nothing at all. Time management and positive thinking are skills you've probably already mastered by now, in view of how much you've already achieved. If you find yourself worrying and predicting some awful consequence, yell "STOP" to yourself and go on. Many times when we imagine the worst, we actually help it happen.

Remember to take time outs, both while on the ground and even when flying. Even a jet aircraft has many relatively slow periods; take advantage of them for brief rest periods. You, similar to the aircraft you will eventually master, can rapidly reach a state of fatigue if you over stress. What would happen if you flew an aircraft at nine (9) G's for extended periods of time?

Recognize that some things can be changed, while others cannot. Although your training and your flying career will likely be rewarding, there are bound to be periods of frustration. Often it is better to change your appraisal of the situation rather than bang your head against the wall trying to change the situation. Take a broader view, it will help you tolerate the valleys as you eagerly anticipate the peaks.

The value of exercise cannot be overemphasized. Flying, although physically demanding at times, is actually a rather sedentary activity. Study is notoriously sedentary. Try to get at least the amount of exercise you enjoyed before you commenced training. Most of you are currently in excellent physical shape. Physical fitness is a side advantage of exercise, an attribute that comes in handy in the flying environment. Make time in your day for some exercise, even if only a walk of thirty minutes. Many students enter flying training and make the mistake of not continuing to exercise.

Sleep problems are the number one indicator of stress in an aviator. Your sleep will compete with many other demands, robbing you of adequate rest. Going from early to late weeks is a reality with which you will contend. Use the techniques listed on the handout (distribute sleep hygiene handout) *to best ensure the best sleep you can get.*

Diaphragmatic or deep breathing is a simple technique that can be mastered with practice. It is especially important for individuals who survive on supplemental oxygen. It is also useful in everyday life. (Hand out diaphragmatic breathing handout and lead a practice).

Above all else, consistently give yourself a pre-flight inspection, just as you pre-flight your aircraft.

Ask yourself:

- *Have I eaten a nutritious meal? Will I get hungry during the flight? Will nutritious meals be available when/where I land?*

- *Do I need to use the restroom? (Sort of like the question you ask a five-year-old child and we all know how that usually goes.)*

- *Have I properly hydrated? (Sounds like it would conflict with the above item.)*

- *Am I mentally prepared to fly?*

- *What do I need to do in the future to feel more prepared for flying?*

- *Is there anything I need to take care of before I leave terra firma?*

- *Have I slept well? (You can't start a trip with a dead battery.)*

Develop your own items as appropriate. Better to ask yourself these question than have the flight surgeon asking your relatives and friends after an accident.

End of talk

Anti-Airsickness Behavioral Checklist (A-B-C)

Pre-Flight

- Practice diaphragmatic breathing (see below).

- Eat breakfast before all flights. Eat lunch before afternoon flights.

- Avoid: Acidic foods (orange or tomato or grapefruit juice), Caffeine (coffee, tea, chocolate milk), Greasy food.

- Drink plenty of water to keep well hydrated.

- Practice good sleep hygiene.

- Don't wear tight underclothes. Keep your flightsuit as comfortable as possible (Don't over-tighten the Velcro fasteners).

During Flight

- Don't forget to breathe! Breathe slightly slower than your instructor pilot (IP). Use diaphragmatic breathing at a slow, comfortable pace.

- Be careful of all head movements. If you need to make a clearing turn, first move your eyes and then follow with your head. <u>Practice</u> this technique during driving and other ground activities.

- If possible, keep two fingertips on the stick at all times when the IP has control of the aircraft. This technique allows you to feel in control and anticipate motion.

- If your feet start feeling cold, wiggle your toes to increase blood flow to the toes.

- If you do get motion sick, whether you vomit or not, don't worry. Many student pilots get motion sick when they are learning to fly and get over the problem with continued flying. If you faithfully use these techniques but continue to get motion sick, your flight surgeon and aerospace psychologist have other programs available. Airsickness is usually the result of stress and lack of adaptation. Many student pilots have used this techniques, coped with airsickness symptoms, and are today successful pilots. Use these techniques even if you're not bothered by airsickness symptoms to enhance your performance.

Diaphragmatic Breathing for Stress Management

- Assume a comfortable posture. Loosen your clothing or flight suit. Loosen your boots or shoes.

- Place your hand lightly on your abdomen.

- Inhale slowly through your nose, allowing the air to go all the way down to your abdomen. Your abdomen should expand, causing your hand to move forward. (Sometimes it is helpful to practice in front of a mirror).

- Hold your breath gently and momentarily.

- Exhale slowly (taking 2-4 times as long to exhale as you do to inhale), pushing your abdomen in with your hand.

- Repeat at a slow, comfortable pace. When flying with an IP, listen to his/her breathing on the "hot mike" and breathe slightly slower than him/her. (This technique can actually keep you calmer than your IP.)

- If you begin to feel dizzy, simply resume your normal pattern of breathing.

This procedure requires PRACTICE, PRACTICE, PRACTICE. Smokers may especially notice a difference in their experience of relaxation using this procedure.

If you determine that they are motivated to fly and are not suffering from manifestations of apprehension, this is a good time to get them started on relaxation and desensitization exercises. Emphasize coping and reassure them. Above all else: Normalize their experience!

Phase 2 (Reassure/Re-Educate)

If a student pilot becomes actively or passively airsick and is referred to the flight surgeon, reassure him/her by normalizing the experience and personally give another copy of A-B-C. Don't just refer to the techniques, give another copy. The aviator probably lost or destroyed the first copy as part of the defense of denial ("I'll never need this!").

Phase 3 (Relax/Desensitize)

If a student continues to experience airsickness, it is time to get a good look at what might be going on. If you are a flight surgeon, this continuing airsickness might also be a good time to make the referral to the local aerospace clinical psychologist, if he/she is available. The aerospace clinical psychologist can plan on spending at least 30 minutes with the student and explore motivation to fly, at what point in the flight they *begin* to feel sick, and their particular symptoms.

Phase 4 (Behavioral Airsickness Management - "BAM")

If the airsickness continues and three active episodes have occurred, it is time to get very aggressive and initiate a more intensive intervention. The program at Sheppard AFB, TX, has been known as the "BAM" program (Behavioral Airsickness Management; Giles and Lochridge, 1985) and involves three consecutive days of spinning the student and coaching how to recognize and deal with symptoms early.

Advise aviator to appear for training with checklist (worn on knee board), wearing flight suit or other appropriate aviation attire and ready to practice some potentially uncomfortable procedures. These procedures will severely test the aviator's motivation. You will be working with them for three consecutive days, so be sure to start in the beginning of the working week. They may not fly for the rest of the day after this training, but may fly before reporting.

Day One

You'll need the ever-vanishing Barany chair or, in a pinch, a substantial swivel chair or even a children's merry go round. Be sure to have plenty of airsickness bags and an available restroom. Prepare some goggles that are blacked out, allowing you to blind fold the aviator. Prepare some items to provide visual stimulation to the aviator, such as a clock with easily movable hands, model aircraft on sticks, mock gauges with adjustable indicators.

When you're spinning the aviator, very obviously follow the suggested procedural checklist provided below. The aviator will take great comfort in the knowledge that you're not just "winging" the procedure. I also promised my late mentor, Dr. G. Kress Lochridge, that I would do all I could to ensure that folks wouldn't do the procedure half-heartedly, see it fail, and then reject the procedure. Practitioners of these techniques afford them almost religious or even cult status. Use these techniques accordingly or risk the consequences.

Brief Aviator about First Ride

- Explain reason for the vision-blocking goggles (to eliminate visual stimulation and to help focus attention on body/symptoms).

- Demonstrate the maneuvers:
    Right and left aileron rolls (held for 30 seconds). Aviator leans entire body.

    Right and left clearing turns. Aviator moves head only.
    Forward loop of dive. Aviator leans entire body forward.

- Explain the 1 – 10 scale:
1-4	Can fly O.K.
5-7	Need to fly straight and level
8-9	Need IP to take the stick
10	Actively airsick

- Emphasize importance to aviator to paying attention to body. Teach them to distinguish between the various symptoms of airsickness that occur before any emesis. These symptoms include light-headedness, sweating, salivation, or symptoms idiosyncratic to that aviator.

First Spin (First Day)

One revolution every two seconds/30 revolutions per minute.

- Have aviator wear blacked-out goggles and have them continually assess themselves on the scale. Ask them if they have any symptoms in the head, stomach, etc.

- Allow aviator two-four minutes to adjust to the motion stimulation (depending on where they are on the scale).

- After two-four minutes (assuming they are no higher than 4 on the scale), have them do three or four maneuvers.

- Effect an abrupt touchdown.

- Allow five – ten minutes for aviator to take a break, splash water on face, clean up, and dispose of airsickness bag (as necessary).

Second Spin (First Day)

- Spin in opposite direction.

- No goggles this ride.

- Ten minutes or until emesis (vomitting) occurs.

- Encourage aviator to look around the room (to demonstrate the effects of visual stimulation).

- Have aviator perform some maneuvers.

- After emesis, have them relax, wash up, and return.

- Give tutorial on diaphragmatic breathing and have them practice. Emphasize that diaphragmatic breathing lowers Sympathetic Nervous System (SNS) arousal.

- Encourage them that they can cope with their symptoms and even gain some control. Possible to feel sick and *not* throw up.

Third Spin (First Day)

- Ten minutes or less.

- Encourage them to practice remaining in control - using diaphragmatic breathing as necessary.

- Have them assess themselves on the 1 - 10 scale and do a maneuver or two as indicated.

- Gradually bring them in for a "landing" (count down as gradually bring them in.

Day Two

- Explain and practice "drop off" technique (Aviator simultaneously contracts all skeletal muscles and holds them for five to ten seconds and then rapidly relaxes these muscles while diaphragmatic breathing [decreases sympathetic arousal]. Instruct aviator to use the full procedure during this training but to modify it to include only those muscles that yield the most relaxation while flying).

First Spin (Second Day)

- Ten minutes.

- Have them do four maneuvers.

- Give the aviator lots of coaching/positive encouragement.

- Remind them to do drop off.

Second Spin (Second Day)

- Ten minutes
  Start visual training
  (Instruct aviator to keep eyes on the nose of the aircraft; pursue presented boogies with their eyes and to form an image in their mind and then read that [iconic memory]. Present the aviator with an inflight emergence and have them read from checklist).

Day Three

- Stroke them on their obvious motivation as evidenced by their willingness to stick with this grueling training.

First Spin (Third Day)

- Ten minutes.

- Have them do four maneuvers – at their own pace.

- Offer no coaching.

Graduation Spin

- Explain that a training aircraft typically falls at the rate of 550 feet per second, so to go from a typical ceiling of 25,000 feet to sea level takes about two minutes.

- Spin at the rate of three minutes at one revolution per second/60 revolutions per minute.

- Straight and level flight (no maneuvers).

- Bring them to an abrupt stop and have them recite, from memory, spin recovery procedure.

- Upon their first solo (if they are a student aviator) they are to give you their class patch. If they are a trained aviator, they are to give you something of aviation significance upon sustained lack of airsickness.

Case Vignette

A student pilot is seen for a work up of suspected vertigo. Physical examination suggests no physical basis for his symptoms, which turn out, were actually more indicative of airsickness. He is therefore referred to a psychologist to assess his motivation to fly. The student pilot states if he is not able to resume flying training duties, it will be "like dying." He explains that his family, including his parents and his wife, supports his flying because they realize how important it is to him. Assessed to have moderate to high motivation to fly, he is taught diaphragmatic breathing, coached on appropriate head movements, and taught stress management techniques. He is also given a five-tape program of flight desensitization, which emphasizes coping in the aviation environment. Additionally, he is exposed to motion stimulation via the Barany chair and taught, in real time, to employ the skills he has been learning. He is returned to his unit with instructions to fly with an instructor pilot for a reasonable number of flights. The student pilot sends word after about five months announcing that he is successfully coping with his airsickness and subsequently reassigned to jet upgrade training, graduating at the top of his class.

12 Consulting to an Aircraft Mishap/Accident Investigation Board

Mishaps *vs.* Accidents

The USAF conducts two investigations after a major loss of life or materiel. The Safety Investigation Board (SIB) terms catastrophic events "mishaps" and is aimed at identifying causes to help prevent similar events in the future. Another board, often termed the "collateral board" or "accident board," is convened to preserve evidence. The SIB is typically composed of individuals from bases other than that of the mishap crew and convenes at the USAF base closest to the mishap site. They conduct a privileged investigation; they are permitted to grant witnesses confidentiality. Furthermore, testimony is not revealed to the accident board or to anyone outside the safety community (which, under penalty of federal law, can only use information for future mishap prevention). Each board has a flight surgeon as a member who may, in consultation with the Board President (who is a rated colonel or higher), request a human factors expert as a consultant.

A Role for Human Factors Consultants

Human factors are responsible for the vast majority of aircraft mishaps, including those cases where a human operator did not initiate the mishap sequence, but also did not intervene to stop it. The USAF is increasingly considering the impact of maintenance factors and the cultural climate of the unit. For example, does the unit emphasize mission completion regardless of the risks involved or is safety the over-riding consideration, often to the detriment of mission completion? Of course most units are somewhere in the middle of these two extremes.

Human factors experts should not underestimate their potential contributions to SIBs. Trained as scientists and often with an appreciation of group process, human factors experts can help the SIB reach valid conclusions. For example, appreciating baserates is a skill not readily demonstrated by many untrained individuals. In other words, just because someone did not (yet) crash, that is not necessarily evidence of superior safety.

Upon finding yourself consulting to a board, you may need to clarify your role as a consultant to the board and educate them as to your skills, and possibly your limitations. Be flexible as boards have various, and sometimes changing, needs for the services of a human factors consultant depending on the personalities of the board members, the dynamics of the group, the extent of human factors involved in the mishap, and the interests and abilities of the flight surgeon member.

It's important to be an excellent consultant, but it's at least equally important to be an outstanding officer (if appropriate) or team member. Pay strict attention to customs and courtesies. Be on time for all board meetings, and arrange with the board president, in advance, if you will be late for, or absent from, a board meeting.

Data Collection

Show up ready to work, but beware! You may be tempted to start offering your expertise immediately. Don't! You'll have that opportunity later. Gather information. Ask questions. The aircrew members of the board do not expect you to know everything about aviation. That is not why they invited you; they know about that part of the mishap. You're there to add your specialized knowledge.

Crash Site

Visit the crash site if you can. Consider visiting the site with a surviving crewmember, the flight surgeon, and perhaps another board member. It may or may not affect your input much, but it will help you see the big picture. When you visit the crash site, don't touch anything. You could destroy fragile evidence or hurt yourself. Hazards include sharp objects, ejection systems, unexploded ordinance, and toxic materials.

Human Remains

You may have the opportunity to attend the autopsy, view photos of the victims at the crash site and at autopsy, and may listen to the cockpit voice recording prior to and during the mishap. Don't underestimate the emotional impact this material may have on you and others. Consider where these items are stored and viewed. Often boards will keep separate collections of photos with and photos without human remains.

Data Analysis

Be very careful to distinguish opinions versus facts. After a while it becomes difficult to distinguish what people hypothesize from what they know. Many boards keep a three-column list labeled:

1.) What we know	2.) What we believe	3.) What we need to find out

Don't forget your training in hypothesis testing, data analysis, and interpretation. Your ability to do simple statistics will go a long way. Don't let the board make inferential errors. Boards are often tempted to draw general conclusions from low baserate events. For example, just because somebody else didn't crash, doesn't mean they were flying safely, and just because something is very unlikely, doesn't mean it didn't happen.

Seeking Consultation

If you reach the limits of your expertise, call someone to ask circumscribed questions. Be sure, however, to clear your questions with the board president or other appropriate authority as they will help you to be discrete and advise you as to what type of confidentiality or privilege statement to read to your consultant. Over time you will develop a network of experts; be sure to keep a record of contact information for future reference. Ask tailored questions of your experts and tactfully insist they not draw conclusions from the questions you have asked nor discuss your conversation with other parties.

Board Dynamics

The group dynamics of a board can be intense. Everyone is under pressure and fatigued, but united in the singular goal of learning what went wrong and initiating the process of fixing it, if possible. Emotions may run high at times, particularly if fatalities are involved. Some boards will respond by avoiding conflict while others can't move beyond it. Help the group deal with the conflict, and watch your own response to conflict. You may be dealing with board members with little appreciation of human factors. This group, however, is very bright. Be diplomatic and extremely patient. You may be most effective if you think of yourself as an educator. (See previous section on "Speaking the Tongue.")

Other Roles

As time evolves, you may become a trusted ally of the board president due to your logic, writing abilities, ability to synthesize and sum up complex situations, and talent for putting together a briefing. Remember, however, that your primary connection to the board is via the flight surgeon. Also, you may be able to use your skills to benefit the group process, but you're not there to be a therapist or formal facilitator. You'll have the opportunity to observe and influence group process, and perhaps individual board members. Board members are exposed to many stressors: the mishap site, the wreckage, human remains or injured aircrew members, photos, cockpit voice recorders, and perhaps the notion that "a good pilot screwed up." Listen to what they have to say about it.

If you're still around at the end of the process, offer to serve as a mock audience member and critically evaluate the board president's briefing. Ask tough questions so that the president can get some practice. Don't underestimate the value of role-play, but do present it as "practice." Remember, aviators respect planning so as to appear "spontaneous."

Questioning the Aircraft Mishap Crew's Colleagues and the Leadership

Introduce yourself to the interviewee. Remember to have them read any advisory statements. Often, effective interviewing is a critical need of the board and witness information may have already been contaminated. Be aware that surviving crew members may never have been given the opportunity to "tell their story" and that your encouragement to do so

·without interruption may lead to valuable new information as well as having therapeutic benefit for them. Interviews should be carefully planned. Identify what information needs to be collected, who needs to be interviewed, what questions need to be asked, who should conduct the interview, and who should attend the interview. Board members may not realize your expertise in conducting effective interviews; show them your stuff. Help board members prepare their interviews (see below) and give a crash course in how to interview effectively. Emphasize rapport building, open-ended questions, follow up questions, and non-leading questions. Suggest that no more than two interviewers sit in on the interviews. The others can listen to the tape. Important interviews can be transcribed and others simply summarized, so taking notes can save time and effort. Ask that key interviews be transcribed. If you've done a key interview alone, suggest that a board member, such as the flight surgeon (if you're a psychologist), listen to these tapes.

Ask for permission to tape. Ensure awareness of difference between safety board and accident board, if such a distinction exists in your jurisdiction (see above). Be sure to start the tape; announce the person you are interviewing and the date. Close the interview on the tape with a statement that this concludes the interview with the person.

Ask which of the crewmembers they knew and worked with and how they came to know them so that you will know which and how many of the crewmembers you will have to inquire about.

An alternative to a structured interview is to start general ("Tell me about...") and then get more and more specific. This strategy requires taking notes, or an exceptionally good memory, as there will be many avenues to pursue. Have them use the airplane models you provide but make sure they describe it for the audiotape.

Background Questions[1]

What is your position in the squadron?

What are your crew qualifications?

How long have you been here?

How long have you been in a flying position either here or in another unit?

What was/were your previous assignment(s)?

What was/is your relationship with (Crewmember)?

How well do/did you know (Crewmember)?

How long have/did you know(n) (Crewmember)?

Have you ever flown with him/her?
      When?
      How often?
      In what position?
      On what types of missions?

Personality and Interpersonal Style

How would you describe Crewmember?

How would you describe his/her personality?

How would you describe his/her relationships with other members of the squadron?

How assertive was he with other squadron and crewmembers?
(For example, did he listen to understand or listen to refute, was he auto-cratic and argumentative or cooperative and team-oriented, was he arrogant and self-absorbed or was he able to listen to and consider other people's opinions?)

Personal Life Events

What kind of work demands did crew member talk about?
    Flying vs. additional duties?

What kind of family and relationship demands did he talk about?

What were his/her goals for the future in the organization and in his/her personal life?

What is your PERSTEMPO (pace of life) like?
      Do you think that's typical?

How are additional duties distributed in your squadron?

What's the leave policy like in this squadron?
    Formal and real world?

What kind of emphasis is placed on professional military education (PME) and off-duty education?

What does it take to get ahead in this organization?

Physiology

What were his/her fitness habits?

Was he/she health conscious?

What were his/her smoking and drinking habits?

How would you describe him/her at squadron parties?

Flying Skills/Airmanship/Systems Knowledge/Habit Patterns

How would you describe his/her flying skills?
    Systems knowledge?

What would you say were his/her greatest strengths as a pilot?
    As an aircraft commander?

What areas would you have suggested he make adjustments or improvements?

What did/does he/she like most about flying?

How would you describe his/her airmanship qualities such as discipline and judgment?

Have you ever experienced an inflight emergency (IFE) or incident while flying with Crewmember?
>What happened?
>How did he and the crew handle it?
>What was the outcome?

How did your crew approach planning and briefing in missions with Crewmember?
>What types of missions were they?
>Any special details such as night, terrain, obstacles?
>What was the outcome?
>How did the crew divide responsibilities?
>How were responsibilities divided between more experienced vs. less experienced crewmembers?

What kind of crewmember is/was Crewmember?
>What was his/her style in running the crew?
>*(Did he take control or delegate? Did he delegate too much?)*
>What was the atmosphere like in the cockpit?
>What was Crew member's style of cockpit communication?
>How did he accept input and provide feedback?
>How would he identify problems or raise issues?
>Would you say that Crewmember was a mission-hacker?
>>In what way?
>On what occasions have you seen him bend the rules or take short cuts?

What was his/her attitude toward the politics of the squadron and the wing?
>What is your attitude toward the politics of the squadron and the wing?

How is this/her squadron/wing different from previous assignments?

Squadron Functioning

What is the squadron's safety policy--formal and real world?

How would you describe the squadron/wing Risk Assessment / Risk Management Program?

What is the general attitude about regulations and standard procedures?

What are the consequences of refusing a mission for the squadron?
> For an individual?

What are the consequences of raising safety concerns?

Have you ever refused a mission?
> What happened when (would happen if) you did?

How do you perceive the pressure to complete a mission?

How would you describe the general level of experience within the squadron/ in the wing?

How would you describe the squadron flight planning facilities; adequacy of pubs/written guidance/charts in the squadron and on the aircraft?

How would you describe the squadron/wing training program related to flying duties?

Do you feel adequately prepared and trained for all missions you might be required to perform?
> Is that because of training or your own personal experience and initiative?

What is the attitude about CRM?

What is the operational (OPS) TEMPO like?
> How does it affect you and your squadron mates?
> How does it compare with past assignments or with a couple of years ago?

Preparation and Briefing

What do you know about the crews' preparation for this particular mission?

What was their attitude about going on this particular mission?

What did he have to say about the other members of the crew?

How would you go about preparing for this mission?

What points of emphasis would be important in planning for the mission?

How would you divide your time?

What pubs/aids would you check, get?

Data Gathering in Mishap Squadron

Timeline of events leading to mishap

Promotion timing

Unfavorable information files (UIFs)

Additional Duty Roster

Squadron Tasking Log

Leave history and policy statement

Finance 6-month look at TDY (temporary duty) for all personnel

30/60/90 totals and last nighttime flight

Accident history of the Wing

Personal Calendar Copy

Authorized/Assigned Personnel

Questions for Leadership

Is there a wing or squadron responsibility to review and/or approve missions before they are assigned to the crew?

What criteria were used to build this crew?

What competing missions were there?

What was the possibility of requesting a change or refusing the tasking?

Which missions are briefed to whom?

What competing priorities were there at the time this mission was being planned?

Manning, Experience, and Training

What is your sense of the experience level of the wing?
> How do you track it?
> Could you show me the tool(s) you use?

Have you raised any concerns about overall manning or experience levels?
> Any documentation of those concerns available?

What's your assessment of quality of your training program?

What would you say are the strengths and opportunities for improvement?

How do you balance safety with the need for mission accomplishment?

What is the status on implementation of Operational Risk Management?

What message do you try to communicate to the squadron?
(For example: Refusal of missions which exceed ability; mentoring of new flyers; personal risk taking/ risk assessment/management)

What are the consequences of refusal of a mission?

What are the consequences of non-compliance with standard operating procedures (SOPs)?

What are the consequences of short cuts?

What are the criteria used to select and build crews?

What is the quality control (QC) process for composition of crews for designated missions?

What is the QC process for mission planning?

Note

1   These questions are tailored to a multiple airframe. Modify them as appropriate, for single-seat airframes by changing "crew" to squadron/flight member and wingman.

Glossary

Psychologists, particularly those new to the aviation arena, often find themselves at a disadvantage due to not understanding the argot/terminology of aviators, particularly military aviators. While it is vital to avoid speaking in psychological jargon ("psychobabble"), it is almost equally important to not use these terms until you are completely certain of their meaning and the context in which they should be employed.

A/A - Air-to-Air

A/G - Air-to-Ground

AAA - "Triple A," Anti-Aircraft Artillery (also flack)

AC - Aircraft Commander

ACE - combat pilot who scored five or more air-to-air kills

ACES II - Advanced Cockpit Escape System (state of the art ejection seat)

ACM - Air Combat Maneuvers

ACMI - Air Combat Maneuvering Instrumentation

ACRO - Acrobatic flight

"ACROSS THE POND" - Across the Atlantic, Pacific, etc.

ACTION - A specific point to start a surface attack, generally 4.5 miles from target (figured on geometry)

ADI - Attitude Direction Indicator

ADIZ - Air Defense Identification Zone

ADVERSARY - An Aircrew or Aircraft flying as an opponent during Air-to-Air training

AGGRESSOR - Pilots trained in enemy tactics therefore assuming the role of "enemy aircraft" during combat exercises

AGL - Above Ground Level

AGM - Air-to-Ground Missile

ANGLE OFF - The angle between the defender's flight path and the attacker's flight path measured from the defender's six o'clock

AOA - Angle of Attack

ASPECT ANGLE - Angle between the defender's longitudinal axis and the attacker's position measured from the defender's six o'clock

ATC - Air Traffic Control

ATO - Air Tasking Order, frag for the first days of war

AT ZERO – Having an enemy fighter (bandit) on your tail (at your six)

AUGER IN – Crashing an airplane, usually at a high angle of attack and high speed (also a popular name for bars/pubs that cater to an aviator crowd)

AWACS - Airborne Warning And Control System (a.k.a. "Ear Wax")

BAILOUT - Eject, prebriefed work (BAILOUT 3) on the third one you leave. Any time ejection needs to be discussed, you use another word.

BANDIT - Known enemy aircraft

BDA - Bomb Damage Assessment

BFM - Basic Fighter Maneuvers

BINGO - Fuel state that prohibits safe continuation of present mission

BLIND - Lost visual contact

BLOW THROUGH - Directive/informative call that indicates aircraft will continue straight ahead at a high rate of speed at the merge and not turn with the target

BOGEY - An aircraft which is unidentified

BREAK TURN - Maximum performance, energy depleting, defensive reaction to a threat in order to live

BREAK (UP/DOWN/RIGHT/LEFT) - Directive call to initiate an immediate break turn in the direction indicated

BUBBLE HEAD - F-16 pilot/driver/jock

BUGOUT - Combat separation with intent to permanently separate particular engagement/attack

BUY THE FARM – Get killed in action

BVR - Beyond Visual Range

CAS - Close Air Support

CELL - Two or more Tanker/Bombers flying in formation

CHAFF - Dispensed out the back to break off enemy radar

CHECK SIX – Watch your tail

CLEAR(ED) - No threats are observed

CLEARED DRY - Not to drop ordinance

CLEARED HOT - To pickle ordinance; peacetime, range control officer (RCO); wartime, forward air controller (FAC)

COMEBACK HIGH/LOW/LEFT/RIGHT - Call requesting the addressed fighter to reposition accordingly

COME OFF LEFT/RIGHT - Direction to turn after attack so that mutual support may be regained

DEFENDER - An aircraft of any type engaged in defensive maneuvering

DEFENSIVE TURN - A basic defensive maneuver designed to prevent an attacker from achieving a launch or firing position. The intensity of the turn is governed by the angle-off, range, and closure of the attacker

DEPLOY - A directive for flight to begin; engaged tactics; positioning will be briefed

DG - Distinguished Graduate (curve buster)

DNIF - Duties Not Including Flying. Medical grounding - no "up slip" (Drunk, not interested in flying)

DSO - Defensive System Officer

DT - Dive Toss

ECM - Electronic Countermeasures

ECCM - Electronic Counter-Countermeasures

EGRESS - To bailout

ELEMENT - A flight of two aircraft

ENGAGED - Indicates fighter or element is maneuvering to attain or deny weapons release parameters or is in visual arena maneuvering in relation to the target. The aircraft/element that's making the bandit predictable

EOD - Explosive Ordinance Disposal

EOR - End-of-Runway

ETA - Estimated Time of Arrival

EW - Electronic Warfare

EWO - Electronic Warfare Officer

EXPEDITE - As quickly as possible (slang: PUSH IT UP)

EXTEND LEFT/RIGHT - Gain energy and distance using proper energy profile with objective of reentering the fight

FAC - Forward Air Controller

FALCON - F-16, "lawn dart," "electric jet," Viper

FCF - Functional Check Flight

FCP - Front Cockpit

FEBA - Forward Edge of the Battle Area

FENCE - Acronym standing for Fire control/Fire power, Emitters, Nav aids, Communications, and ECM. Also the boundary separating hostile and friendly forces

FEZ - Fighter Engagement Zone

FLARE - Dispensed out the back to fool any IR missile

FLIR - Forward Looking Infrared

FLOT - Forward Line of Troops

FOD - Foreign Object Damage

FRAG - Short for fragment (Tasking order for wing or squadron to go fight (i.e. "as fragged" is doing what you're told)

FSO - Flight Surgeon's Office (place to avoid)
        - Flying Safety Officer

FUR BALLS – Confusion during a multiple aircraft dogfight

GCI - Ground Control Intercept - Radar controls

GLOC - G induced Loss of Consciousness

GOO - Bad weather

"GIVE IT BACK TO THE TAX PAYERS" - Eject

GRAPE - Non-maneuvering (due to stupidity or lack of energy) Bandit

GS - Ground Speed

GUNS - Call indicating simulated gun employment (Navy)

HAHO - High Altitude High Opening

HALO - High Altitude Low Opening

HARD TURN - Maximum turn that will *sustain energy state*

HAWK - Staying above the fight - not engaging
        - British fighter aircraft

HIGH ANGLE GUNSHOT - Gunshot at high aspect angle where tracking cannot be maintained. Also referred to as a "snap shot"

HOSTILE - A contact or aircraft positively identified as an enemy in accordance with (IAW) operational rules of engagement (ROE)

HOUND DOG - Informative call from wingman to leader requesting a role change from supporting to engaged. It means I am tally ho, visual, and in an advantageous position to engage

HSI - Horizontal Situation Indicator

HUD - Head(s) Up Display

IAW - In accordance with

ICS - Intercom System

IFF/SIF - Identification Friend or Foe / Selective Identification Feature

IFR - Instrument Flight Rules

ILS - Instrument Landing System

IMC - Instrument Meteorological Conditions (Goo/Soup)

"IN THE SOUP" - Flying in the weather

INS - Inertial Navigation System

IP - Instructor Pilot (Not necessarily "Identified Patient")

IR - Infrared

JAAT - Joint Air Attack Team

JAFO - Just Another F------ Observer

JINK - Unpredictable maneuvers designed to spoil an opponent's gun tracking solution

JOKER - Prebriefed fuel state above Bingo, usually where separation will be started

KCAS - Knots Calibrated Airspeed

KIAS - Knots Indicated Airspeed

KTAS - Knots True Airspeed

KICK-HIM-OUT - A request for an aircraft being attacked to increase angle off/aspect angle on the attacker. Generally used internally by two seat crews to maintain tally

KNOCK-IT-OFF - Terminate any fighting maneuvers now in progress

LANA - Low Altitude Night Attack

LANTIRN - Low Altitude, Night Infrared Navigation System

LATF - Low Altitude Tactical Formation

LATN - Low Altitude Tactical Navigation: Low altitude training using the fundamental aspects of dead reckoning and point-to-point low altitude navigation with or without prior route planning

LEAD PURSUIT - The attacker's instantaneous flight path vector or nose is pointed in front of the defender

LETHAL ENVELOPE - The envelope within which parameters can be met for successful employment of a weapon

LIFE VECTOR - The plane of total G vector. It extends through the center of the aircraft perpendicular to the wings

LIFT - Lead-In Fighter Training

LOS RATE - Line-of-Sight Rate

LOWAT - Low-Altitude Air-to-Air Training

LUFBERY - A circular, stagnated fight with nobody having an advantage

MDA - Minimum Descent Altitude

MERGE - Radar returns have come together or the meeting of two or more aircraft at high aspect

MIL - Military Power

MILK RUN – An uneventful, easy combat flight

MQT - Mission Qualification Training

MR - Mission Ready

MS - Mission Support

MSA - Minimum Safe Altitude

MSL - Mean Sea Level

NM - Nautical miles

NO JOY - No visual contact with the bogey/bandit. Opposite of tally ho

NOTAMS - Notice to Airmen (Notices issued to aircrews on what they might encounter en route)

NVG - Night Vision Goggles

OAA - Out And About

OFF HIGH/LOW/LEFT/RIGHT - Attack is terminated or completed and aircraft is repositioning as stated

OSO - Offensive System Officer

OTL - Out to Lunch

OVERSHOOT - Attacker is forced outside defender's flight path, or in front of his 3/9 line, or both

P^3 - Piss-poor protoplasm, an aviator who doesn't fit in

PADLOCK - I have visual/tally-ho and cannot look away from the fighter/target without risking loss of visual contact

PAX - Passengers

PICKLE - Release ordinance

PIREPS - Pilot Inflight Reports

PIROUETTE - Vertical repositioning maneuver

PITCHBACK LEFT/RIGHT - Informative call to execute a nose high heading reversal to reposition as stated

PK - Probability of Kill

PLAYERS - All participating aircraft in the engagement (friendly and adversary)

POP - To start up to delivery parameters

POPEYE - Flying in clouds or area of reduced visibility

POSITION/SAY POSIT - A question that asks "where are you?"; usually a flight member asking "where are you in relation to me?"

PRESS - Continue to attack, I have all players in sight and I am in a position to support

PUCKER FACTOR – Level of anxiety experienced by an aircrew

PULLDOWN - Altitude to start roll in

PUNCH OUT – Eject, also going home for the evening, moving to a new duty station, retiring

RECCE - Reconnaissance aircraft

RF - Radio Frequency

ROE - Rules of Engagement

ROGER - I have received your transmission. Does not indicate compliance (shortened - ROG)

RON - Remain Over Night (Usually a good thing)

ROZ - Restricted Operating Zone

RTB - Return to Base

RTU - Replacement Training Unit

SAFE AREA - Selected Area for Evasion

SAM - Surface to Air Missile

SANDWICH - A situation where an aircraft/element finds itself in between opposing aircraft/element(s)

SAR - Search and Rescue

SAT - Surface Air Tactics

SCAR - Strike Control and Reconnaissance

SCRAMBLE ORDER - Command authorization for tactical flight establishing an immediate departure time

SD - Spatial Disorientation

SEAD - Suppression of Enemy Air Defenses

SEPARATE - Leave the fight/engagement due to loss of advantage, change of odds or situation. Similar to bugout except not necessarily permanent

SEQUENTIAL ATTACK - Swapping of roles of engaged and supportive fighters as one or the other comes into a more favorable position to achieve a kill

SHACK - On range - a hit w/in 27 feet (also means absolutely correct, a term of agreement)

SHOOTER - Aircraft designated to commit air-to-air ordinance

SID - Standard Instrument Departure

"SIERRA HOTEL" - Sh-t hot, good job, good deal

SLICEBACK - An optimum to maximum performance turn with the nose of the aircraft below the horizon, reversing flight path direction while maintaining maneuvering energy

SNAP SHOT - High angle-off or passing gunshot - not long enough to track (call "SNAP" on the radio)

SOBs - Souls On Board

SOF - Supervisor of Flying

SOP - Standard Operating Procedure

SPIT OUT - An unintentional exit of the engagement

SPLASH - Missile time-of-flight is expired or target destroyed on bomb impact

SPLIT - Entities described are separated/separating; directive to maneuver with separate targets

STATUS - Inquiry as to partner's perceived tactical situation; response will be "neutral, offensive, or defensive," as appropriate

STOP - Strategic Orbit Point

SUPPORTING - Fighter supporting engaged fighter
SWITCH - To break off an attack on one enemy in favor of attacking another

TA - Training Area

TALLY HO - Visual sighting of a target. Opposite of No Joy

"TANGO UNIFORM" - "Tits up," gone bad, broken

TARF - Trainer, Attack, Reconnaissance, Fighter

TARGET-RICH ENVIRONMENT – More targets than you have ordinance

TDY - Temporary Duty (away from permanent station, usually a good thing)

TGT - Target (on the ground usually)
　　- Means change (always a triangle; i.e. "don't do something stupid and frag yourself"

TOT - Time on Target

TR(s) - Training Rule(s)

TRACKING - Act of maintaining aiming index on an aerial target while employing the gun

TSEM - Turning Safe Escape Maneuver (to avoid frag and stay low out of harm's way)

TTB - Tanker, Transport, Bomber

TTG - Time to Go

TUMBLEWEED - No tally, no visual, no clue

TURN AND BURN – To service an aircraft quickly and get it airborne again (a.k.a. hot turn)

TX - Training

UNABLE - Cannot comply

UNKNOWN - Information not available; for example, an unidentified target

UPT/UHT/UNT - Undergraduate pilot/helicopter/navigator training

VFR - Visual Flight Rules

VID - Visual Identification

VISUAL - Visual sighting of a friendly participant. Opposite of blind

VMC - Visual Meteorological Conditions

WC - Weapons Controller

WILCO - Will comply

WINCHESTER - No armament remaining

WIRE - Release angle (i.e. 10, 20, 30, 45)

WIZZO – Weapons systems officer, also known as backseaters, GIBs (Guys In the Backseat), trained bears

WOXOF - Weather obscured, zero visibility

WX - Weather

ZOOMIE - "Grad," graduate of
"The Zoo" (USAF Academy)

References

Adams, R. R. & Jones, D. R. (1987). The Healthy Motivation to Fly: No Psychiatric Diagnosis. *Aviation, Space, and Environmental Medicine, 58*, 350-354.

American Psychiatric Association. (1991). Diagnostic and statistical manual of mental disorders. (4th ed.). Washington, DC: Author.

Ashman, A. & Telfer, R. (1983) Personality profiles of pilots. *Aviation, Space, and Environmental Medicine, 54*, 940-943.

Brown, J. D. & McGill, K. L. (1989). The cost of good fortune: when positive life events produce negative health consequences. *Journal of Personality and Social Psychology, 57*, 1103-1110.

Butcher, J. N. (1980). The role of crisis intervention in an airport disaster plan. *Aviation, Space, and Environmental Medicine, 51*, 1260-1262.

Carretta, T. R., Retzlaff, P. D., Callister, J. D., & King, R. E. (1998). A comparison of two U.S. Air Force Pilot Aptitude Tests. *Aviation, Space, and Environmental Medicine, 69*, 931-935.

Chidester, T. R., Helmreich, R. L., Gregorich, S. E., & Geis, C. E. (1991). Pilot personality and crew coordination: Implications for training and selection. *The International Journal of Aviation Psychology, 1*, 25-44.

Costa, P. T., & McCrae, R. R. (1992). *Professional manual: Revised NEO Personality Inventory (NEO PI-R) and NEO Five-Factor Inventory (NEO-FFI)*. Odessa, FL: Psychological Assessment Resources, Inc.

Dockeray, F. C. & Isaacs S. (1921). Psychological research in aviation in Italy, France, England, and the American Expeditionary Forces. *Journal of Comparative Psychology, 1*, 115-148.

Ewing, J. A. (1984). Detecting alcoholism: The CAGE Questionnaire. *Journal of the American Medical Association, 252*, 1905-1907.

Fine, P. M., & Hartman, B. O. (1968). *Psychiatric strengths and weaknesses of typical Air Force pilots*, SAM-TR-68-121.

Folstein, M. D. & Luria, R. (1973). Reliability, validity and clinical application of the visual analogue mood scale. *Psychophysiology, 3*, 479-486.

Giles, D. A., & Lochridge, G. K. (1985). Behavioral airsickness management program for student pilots. *Aviation, Space, and Environmental Medicine, 56*, 991-994.

Haakonson, N. H. (1980). Investigation of life changes as a contributing factor in aircraft accidents: a prospectus. *Aviation, Space, and Environmental Medicine, 51*, 981-988.

Halpern, D. F. (1992). *Sex differences in cognitive abilities* (2nd ed.). Hillsdale, New Jersey: Lawrence Erlbaum Associates.

Helmreich, R. L., Sawin, L. L., & Carsrud, A. L. (1986). The honeymoon effect in job performance: Temporal increases in the predictive power of achievement motivation. *Journal of Applied Psychology, 71*, 185-188.

Hoddes, E., Zarcone, V., Smythe, H., Phillips, R., & Dement, W. C. (1975). Quantification of Sleepiness : A new approach. *Psychophysiology, 10*, 431-436.

Houston, R. C. (1988). Pilot personnel selection. In SG Cole & RG Demarae (Eds.), *Applications of Interactionist Psychology: Essays in Honor of Saul B. Sells.* Hillsdale, NJ: Lawrence Erlbaum.

Hunter, D. R. & Burke, E. F. (1995). *Handbook of Pilot Selection.* Brookfield, VT: Ashgate.

Jackson, D. N. (1984). *Multidimensional Aptitude Battery manual.* Ontario, Canada: Research Psychologists Press, Inc.

Jones, D. R. (1983). Psychiatric assessment of female fliers at the U.S. Air Force School of Aerospace Medicine (USAFSAM), *Aviation, Space, and Environmental Medicine, 54*, 929-931.

Jones, D. R. (1985). Secondary disaster victims: the emotional effects of recovering and identifying human remains. *American Journal of Psychiatry, 142*, 303-307.

Kay, G. G. (1995). *CogScreen Aeromedical Edition professional manual.* Odessa, FL: Psychological Assessment Resources, Inc.

King, R. E. & Lochridge, G. K. (1991). Flight psychology at Sheppard Air Force Base. *Aviation, Space, and Environmental Medicine, 62*, 1185-1188.

King, R. E. (1994). Assessing aviators for personality pathology with the Millon Clinical Multiaxial Inventory (MCMI). *Aviation, Space, and Environmental Medicine, 65*, 227-231.

King, R. E. & Flynn, C. F. (1995). Defining and measuring the "right stuff:" Neuropsychiatrically Enhanced Flight Screening (N-EFS). *Aviation, Space, and Environmental Medicine, 66*, 951-956.

King, R. E., McGlohn, S. E., & Retzlaff, P. D. (1997). Female United States Air Force pilot personality: The new right stuff. *Military Medicine, 162*, 695-697.

Klein, K. E., Wegmann, H. M., Athanassenas, G., Hohlweck, H., & Kuklinski, P. (1976). Air operations and circadian performance rhythms. *Aviation, Space, and Environmental Medicine, 47*, 221-230.

Lyons, T. J. (1991). *Women in the military cockpit,* AL-TR-1991-0068. Washington, DC: U.S. Government Printing Office.

McGlohn, S. E., King, R. E., Butler, J. W., & Retzlaff, P. D. (1997). Female United States Air Force (USAF) pilots: Themes, challenges, and possible solutions. *Aviation, Space, and Environmental Medicine, 68*, 132-136.

McNair D., Lorr, M., & Droppelman, L. (1975). *EDITS manual for the Profile of Mood States.* San Diego: Educational and Industrial Testing Service.

Mills, J. G. & Jones, D. R. (1984). The adaptability rating for military aeronautics: an historical perspective of a continuing problem. *Aviation, Space, and Environmental Medicine, 55*, 558-562.

Novello, J. R., & Youssef, Z. I. (1974). Psycho-Social studies in general aviation: II. Personality profile of female pilots. *Aerospace Medicine*, 45, 630-633.

Picano, J. J. (1991). Personality types among experienced military pilots. *Aviation, Space, and Environmental Medicine* 1991, 62:517-520.

Reinhart, R. F. (1970). The outstanding jet pilot. *American Journal of Psychiatry, Dec*, 32-36.

Retzlaff, P. D., Callister, J. D., & King, R. E. (1997). The Armstrong Laboratory Aviation Personality Survey (ALAPS): Norming and cross-validation (AL/AO-TR-1997-0099). Washington, DC: U.S. Government Printing Office.

Retzlaff, P. D., King, R. E., & Callister, J. D. (1995a). *Comparison of a computerized version of the paper/pencil Version of the Multidimensional Aptitude Battery (MAB)* (AL/AO TR-1995-0121) Washington, DC: U.S. Government Printing Office.

Retzlaff, P. D., King, R. E., & Callister, J. D. (1995b). *U.S. Air Force pilot training completion and retention: A ten year follow-up on psychological testing* (AL/AO TR-1995-0124) Washington, DC:U.S. Government Printing Office.

Retzlaff, P. D., King, R. E., Marsh, R. W., & French, J. (1997). The development of the Sustained Operations Assessment Profile (SOAP) (AL/AO-TR-1997-0094) Washington, DC: U.S. Government Printing Office.

Retzlaff, P. D., King, R. E., McGlohn, S. E., & Callister, J. D. (1996). *The Development of the Armstrong Laboratory Aviation Personality Survey (ALAPS)* (AL/AO TR-1996-0108) Washington, DC: U.S. Government Printing Office.

Retzlaff, P. D. & Gibertini, M. (1987). Air Force pilot personality: hard data on the "right stuff." *Multivariate Behavioral Research, 22*:383-399.

Rippon, T. S., & Manuel, E. G. (1918). The essential characteristics of successful and unsuccessful aviators. *The Lancet, September,* 411-415.

Santy, P. A. (1994). *Choosing the right stuff: the psychological selection of astronauts and cosmonauts*. Westport, Connecticut: Praeger.

Schnieder, D. (1985). Training high performance skills: Fallacies and guidelines. *Human Factors, 27,* 285-300.

Selye, H. (1978). *The stress of life*. NY: McGraw Hill.

Siem, F. M. (1990). Comparison of male and female USAF pilot candidates. *AGARD Symposium on Recruiting, Selection, Training, and Military Operations of Female Aircrew*. Tours, France.

Siem, F. M., & Murray, M. W. (1994). Personality factors affecting pilot combat performance: A preliminary investigation. *Aviation, Space, and Environmental Medicine, 65*, A45-48.

Strongin, T. S. (1987). A historical review of the fear of flying among aircrew. *Aviation, Space, and Environmental Medicine, 58*, 263-267.

Verdone, R. D., Sipes, W., Miles, R. (1993). Current trends in the usage of the adaptability rating for military aviation (ARMA) among USAF flight surgeons. *Aviation, Space, and Environmental Medicine, 64,* 1086-1093.

War Department (1940). *Outline of neuropsychiatry in aviation medicine,* Technical Manual 8-325. Washington, DC.

Wolfe T. (1980). The right stuff. NY: Bantam Books.

Yerkes, R. M. & Dodson, J. D. (1908). The relation of strength of stimulus to rapidity of habit-formation. *Journal of Comparative Neurology and Psychology,* 459-482.

About the Author

Dr. Raymond E. King is a licensed psychologist and native of New Jersey, United States. He received his BA from Rutgers College (New Brunswick, NJ), his MA from Fairleigh Dickinson University (Madison, NJ), and his doctorate from the Illinois School of Professional Psychology (Chicago, IL). During his United States Air Force career, he implemented treatment programs to aid fledgling pilots cope with airsickness and other adaptational and stress responses to the demands of flight. He has taught mishap investigation techniques to psychologists, physiologists, flight surgeons and other human factors consultants. He has served as a psychiatric evaluator to numerous astronaut selection cycles at NASA Johnson Space Center, Houston, Texas. He served as principal investigator on two grants investigating the stressors, career goals, and personality/cognitive characteristics of male and female aviators. He served as the Chief of the Collaborative Systems Technology Branch of the Crew System Interface Division, Human Effectiveness Directorate, during the merger of the Armstrong and Wright Laboratories into the Air Force Research Laboratory. Dr. King is married to the former Krystyna M. Osinski of Cleveland, OH. They have an elementary-school-aged child, Elliott, who is undecided whether to be an astronaut or a "mad scientist" ("just like daddy").

Organizations to Join (or to at Least Know about)

(No endorsement by the United States Air Force or Department of Defense implied.)

(All organizations are located in the US unless otherwise noted and current at press time.)

Aerospace Medical Association (AsMA)
320 South Henry Street
Alexandria, VA 22314-3579
1-703-739-2240
http://www.asma.org

Aerospace Human Factors Association
(A constituent organization of AsMA)
1804 Woodland Road
Edmond, OK 73013
(405) 954-6297

Association of Aviation Psychologists
Department of Psychology
San Francisco State University
1600 Holloway Ave.
San Francisco, CA 94132
1-415-338-1059
http://userwww.sfsu.edu/~kmosier/aap.html
(Membership, which is relatively low cost, includes a subscription to the International Journal of Aviation Psychology)

Australian Association of Aviation Psychologists
PO Box 99
Clifton Beach 4879 Australia
+61 3 9645 5473
http://www.vicnet.net.au/~aavpa/

Division of Applied Experimental Psychology
(Division 21 of the American Psychological Association)
c/o David J. Schroeder
FAA Civil Aeromedical Institute (AAM-500)
P.O. Box 25082
Oklahoma City, OK 73125

European Association of Aviation Psychology (EAAP)
Morgen 23
2408 RK Alphen Aan Den Rijn
The Netherlands
+31-172426479
http://www.eaap.com/

Human Factors and Ergonomics Society
P.O. Box 1369
Santa Monica, CA 90406-1369
(310) 394-1811
http://hfes.org

The Society for Human Performance in Extreme Environments
6052 Wilmington Pike, No. 166
Dayton, Ohio 45459
1-500-447-HPEE (4733)
http://www.hpee.org

The following organization is not one to join; it is a resource
available for consultation. They also publish an excellent
newsletter, *Gateway*, and periodic State of the Art Reports
(SOARs):

Crew Systems Ergonomics Information Analysis Center
(CSERIAC)
(A DoD Information Analysis Center technically managed by the
Air Force Research Laboratory Human Effectiveness Directorate
and operated by Booz-Allen & Hamilton, McLean, VA)

2261 Monahan Way, Bldg. 196
Wright-Patterson AFB, OH 45433-7022
http://www.cseriac.flight.wpafb.af.mil

Printed and bound by CPI Group (UK) Ltd, Croydon, CR0 4YY

21/10/2024

01777084-0019